DIGITAL
SOUND
STUDIES

———

DIGITAL SOUND STUDIES

EDITED BY

Mary Caton Lingold,

Darren Mueller,

and Whitney Trettien

DUKE UNIVERSITY PRESS · Durham and London · 2018

© 2018 DUKE UNIVERSITY PRESS
All rights reserved

Designed by Matthew Tauch
Typeset in Quadraat by BW&A Books, Inc.
Library of Congress Cataloging-in-Publication Data
Names: Lingold, Mary Caton, [date] editor. |
Mueller, Darren, [date] editor. | Trettien, Whitney Anne, editor.
Title: Digital sound studies / edited by Mary Caton Lingold, Darren
Mueller, and Whitney Trettien.
Description: Durham : Duke University Press, 2018. |
Includes bibliographical references and index.
Identifiers:
LCCN 2018003694 (print) | LCCN 2018008824 (ebook) |
ISBN 9780822371991 (ebook) |
ISBN 9780822370482 (hardcover) |
ISBN 9780822370604 (pbk.)
Subjects: LCSH: Digital humanities. | Sound—Recording and
reproducing—Digital techniques. | Sound. | Sound in mass media.
Classification: LCC AZ105 (ebook) | LCC AZ 105 .D55 2018 (print) |
DDC 001.30285—dc23
LC record available at https://lccn.loc.gov/2018003694

Cover art: Heather Gordon, *Frequencies of the Equal Tempered Scale*,
2017. 30" × 30". Oil on canvas. Photo by Alex Maness

CONTENTS

vii		PREFACE
xi		ACKNOWLEDGMENTS
1		Introduction
		Mary Caton Lingold, Darren Mueller, and Whitney Trettien

I THEORIES AND GENEALOGIES

29	1	Ethnodigital Sonics and the Historical Imagination
		Richard Cullen Rath
47	2	Performing Zora: Critical Ethnography, Digital Sound, and Not Forgetting
		Myron M. Beasley
64	3	Rhetorical Folkness: Reanimating Walter J. Ong in the Pursuit of Digital Humanity
		Jonathan W. Stone

II DIGITAL COMMUNITIES

83	4	The Pleasure (Is) Principle: *Sounding Out!* and the Digitizing of Community
		Aaron Trammell, Jennifer Lynn Stoever, and Liana Silva
120	5	Becoming OutKasted: Archiving Contemporary Black Southernness in a Digital Age
		Regina N. Bradley

130	6	Reprogramming Sounds of Learning: Pedagogical Experiments with Critical Making and Community-Based Ethnography
		W. F. Umi Hsu

III DISCIPLINARY TRANSLATIONS

155	7	Word. Spoken. Articulating the Voice for High Performance Sound Technologies for Access and Scholarhip (HiPSTAS)
		Tanya E. Clement
178	8	"A Foreign Sound to Your Ear": Digital Image Sonification for Historical Interpretation
		Michael J. Kramer
215	9	Augmenting Musical Arguments: Interdisciplinary Publishing Platforms and Augmented Notes
		Joanna Swafford

IV POINTS FORWARD

231	10	Digital Approaches to Historical Acoustemologies: Replication and Reenactment
		Rebecca Dowd Geoffroy-Schwinden
250	11	Sound Practices for Digital Humanities
		Steph Ceraso
267		AFTERWORD. Demands of Duration: The Futures of Digital Sound Scholarship
		Jonathan Sterne, with Mary Caton Lingold, Darren Mueller, and Whitney Trettien
285		CONTRIBUTORS
291		INDEX

PREFACE

MARY CATON LINGOLD, DARREN MUELLER,
AND WHITNEY TRETTIEN

It is not at all lost on us, the irony of writing a book about the importance of digital scholarship. The truth is that the journey that brought us—the editors—here began with a simple question: How can scholars write about sound *in sound*? We sensed that the digital turn was an auspicious opportunity for sonic scholarship—that now, at last, when interrogating matters of the audible world it would be easier to incorporate sounds themselves into academic argumentation. We imagined multisensory web interfaces that would seamlessly embed audio into writing, open access databases full of recordings, and experimental sound pieces distributed effortlessly across social media. We wanted to breach the cultural impasse and give sound center stage in an intellectually rich digital space.

These were pipe dreams, but they were also possibilities that we set out to pursue in our own creative-critical work. Our big questions led us to the proverbial toolshed, where we tried to produce something approaching the visionary potential of what we eventually came to call digital sound studies. As is often the case, when we looked around to see what other work was being done in this vein, we discovered a great deal of innovation occurring across multiple fields and in different types of institutions. Scholars were composing apps for playing with sound, designing signal processors, publishing podcasts, and creating scholarly communities online. Eager to bring this work into the conversation, we began by building a digital home for experimental scholarship. We solicited provocative work in digital sound studies to be part of a custom-built web collection entitled *Provoke! Digital Sound Studies*.[1]

Since then, each of us has gone on to produce different kinds of multimodal scholarship. With some input from others, including Darren, Mary

Caton founded the Sonic Dictionary, a digital collection of audio recordings created and curated by students in sound studies courses across institutions. She has also collaborated on Musical Passage: A Voyage to 1688 Jamaica, which tells the story of some of her research in the form of a digital, audio-rich custom website. Whitney is writing a hybrid print/digital book and has engineered an innovative journal, *thresholds*, that presses further against the boundaries of scholarly writing and collaboration. At the Eastman School of Music, Darren founded the Media, Sound, and Culture Lab as a site for faculty and students to explore digital sound scholarship, digital pedagogy, and various forms of creative scholarship. An ethos of collaboration is essential to each of these developing projects, a legacy of our work together that initially began in 2011. For us, working closely across disciplinary boundaries remains key to the advance of scholarship on and off the page.

In fact, it was our work on *Provoke!* that revealed to us the necessity of this book. Each medium offers its own capacities—affordances, in the digital parlance—and individual projects by themselves only tell part of the story. A great deal of intellectual labor went into the composition and creation of *Provoke!* and the individual projects featured there. To substantiate and make legible the productive modes of thought driving digital sound studies, a much broader, more in-depth conversation needed to take place. This is just the sort of thing that books—and especially collections that feature multiple voices bound together with a single vision—can accomplish. Digital humanities praxis is made possible by the modes of intellectual inquiry and argumentation that humanists are well trained for. But this book shows that the opposite is also true: that born-digital scholarship generates rigorous intellectual inquiry of the sort well suited to the long-form essay. Identifying and bearing out the fruits of the deeply entangled relationship between these two forms of composition is what this collection sets out to do.

Sound productively unsettles many of our ingrained assumptions about the limitations and possibilities of both print and digital authorship. For instance, books can seem frustratingly silent, but as any good reader knows, they can communicate sound quite effectively. In contrast, digital media feels like a natural home for audio, but designing for sound on the web can be challenging. This is why we need to bring the insights of sound studies to bear upon the emergent field of digital humanities. By provoking both fields toward an experimental and soundful engagement with one another, we envision *digital sound studies* will become an interdiscipline born at the intersection of analysis and innovation.

The contributors in this book practice critical listening to reveal the role of sonic life in digital spaces. They also model how to use digital methods both to enhance the study of auditory culture and, literally, to amplify sound in the academy. This book is therefore for sound studies scholars who wish to understand how digital humanities methods might enhance their own research, and it is for digital humanists who seek to enrich their work with sound. It is also for students and scholars across disciplines who are struggling to make sense of the digital turn and its impact on scholarship, the classroom, and wider publics. To seize this moment is to embark on a great experiment, one with upsides and downsides. Not all digital sound scholarship will be transformative. But by being provocative—by giving *voice* to thought—digital sound studies creates the possibility for new kinds of understanding that can do justice to forms of sonic knowledge: the ancient, the fledgling, the yet-to-be imagined.

NOTE

1 *Provoke!* can be accessed at http://soundboxproject.com, or https://doi.org/10.7924/G8H12ZXR.

ACKNOWLEDGMENTS

MARY CATON LINGOLD, DARREN MUELLER,
AND WHITNEY TRETTIEN

This collection is an ensemble piece. It began as a collaboration between three graduate students, supported by Duke University's PhD Lab in Digital Knowledge and the Franklin Humanities Institute; it comes to an end with twenty-six additional digital sound studies coconspirators across two published collections, digital and print. To those who have entrusted us with sharing their work—Salvador Barajas, Myron Beasley, Danah Bella, Regina Bradley, Liz Canfield, Steph Ceraso, Tanya Clement, Rebecca Geoffroy-Schwinden, Alex Gomez, Kevin Gotkin, W. F. Umi Hsu, Michael Kramer, Corrina Laughlin, Andrew McGraw, Robert Peterson, John Priestley, Richard Cullen Rath, Aaron Shapiro, Liana Silva, Jonathan Sterne, Kenneth Stewart, Jennifer Stoever, Jon Stone, Joanna Swafford, Aaron Trammel, Jonathan Zorn—we offer our heartfelt appreciation and thanks.

This collection would have been impossible without the many individuals who read our writing, helped design our website, navigated tricky issues of digital archiving on our behalf, provided technical support or feedback, or simply offered their friendship and encouragement. They are, in alphabetical order: Dan Anderson, Elizabeth Ault, Ian Baucom, Allison Belan, David Bell, Rachel Bergmann, Nicholas Bruns, Chris Chia, Ashon Crawley, Cathy Davidson, D. Edward Davis, Laurent Dubois, Craig Eley, Dave Garner, Monica Hairston-O'Connell, Marc Harkness, Guo-Juin Hong, Robin James, Meredith Kahn, Paolo Mangiafico, Charles Mangin, Jonna McKone, Louise Meintjes, Liz Milewicz, Brian Norberg, Dan Partridge, Jentery Sayers, Will Shaw, Kevin Smith, Amanda Starling Gould, Jonathan Sterne, Stewart Varner, Jacqueline Waeber, Priscilla Wald, and Mary Williams. We also wish to express our deepest thanks to Elizabeth Ault, Ken Wissoker, and the entire team at Duke University Press for their unflinching support and belief in this project.

Finally, we would like to thank our partners Eric Olsten, Phil Torres, and Stephanie Westen for sharing our frustrations and joys throughout this long process.

INTRODUCTION

MARY CATON LINGOLD, DARREN MUELLER,
AND WHITNEY TRETTIEN

Cats meow over the whir of cars passing by. A grainy shuffling, barely distinguishable from the hiss of the tape, echoes in an apartment before two distinct thumps overwhelm the mix. A floor creaks in the distance; a whistling sigh sounds as a bus driver lifts a foot from the brake.

It was fall 2011 and the three of us were crowded around a laptop, listening. The recording we heard came from the Jazz Loft Project, a collection of digitized audio captured by photojournalist W. Eugene Smith between 1957 and 1965. An obsessive sound collector, Smith left his reel-to-reel recorder running nonstop in his rundown New York City loft. Offering more than four thousand hours of audio, the collection is prized for including rare jam sessions with jazz greats like Thelonious Monk, Sonny Rollins, and Charles Mingus. In addition to documenting an iconic era in jazz, Smith recorded all kinds of ephemeral sounds: snippets of phone conversations, fragments of radio and television broadcasts, the roar of buses driving past the loft. This important collection of reel-to-reel tapes was recently digitized and housed on 5,087 CDS thanks to the work of documentarian Sam Stephenson. We wanted to learn more about the process of digitizing a massive collection of audio recordings, so we were meeting with the archive's cataloger, Dan Partridge, who had just played us the clip.[1]

"It took me weeks," he admitted, "but I finally figured out what those thumps are. It's Smith's cats, playing with the microphone." Dan spent his days in a quiet basement, his ears locked under headphones, listening to the recordings on a computer. As he listened, scrubbing the audio back and forth to hone in on particular noises, his ears became attuned to what he was hearing, and he began to develop a mental map of the acoustic space in Smith's loft. Eventually he could interpret sounds that would be unintelligible to a casual listener—understanding indistinct commotion, for instance, as a cat jumping onto a table. Once he had identified the content of a recording, Dan would scribble down his observations on paper. These handwritten notes were then logged in a spreadsheet. Dan's descriptions are now part of the collection's finding aid and thus render an impenetrably large amount of audio data accessible to researchers.

If we were asked to point to a project that demonstrates the potential of digital media to improve sound-based research, we might well suggest the Jazz Loft Project. Yet, as we learned that day in the basement, nothing about realizing the transformative potential of digital scholarship is as straightforward as it might seem.

Take, for instance, the very notion of "digital media." Sitting between a cabinet of CDs, a box of reel-to-reel tapes, a pile of handwritten notes, and a computer screen displaying a spreadsheet, we confronted a tangle of technologies knit together in ways far more complex than the simple modifier "digital" would indicate. Dan was listening to CDs that store digitized copies of Smith's original reel-to-reel recordings, but since each format encodes sonic data differently, the timestamps on the tapes do not correspond precisely to those on the CDs; what is halfway through the first reel may come at the beginning of the fourth CD, for instance. Moreover, even though the audio data on the CD is "digital," it was at that point still locked on physical media in a basement cabinet. Listening to a particular sound would require finding not only the right CD but a CD player—an increasingly rare bit of technology. While in theory, then, digital copies are more manipulable and "spreadable" than their analog counterparts, in practice they are no more accessible to the average listener than reel-to-reel tapes. From the researcher's perspective, this shift from one platform to another currently signals little more than a loss in fidelity for the Jazz Loft Project.

It is, of course, technologically possible to rip data from the CDs and post the clips online for streaming, assuming one has access to the right software and a server. Yet, again, what is technologically possible is not so easily realized in messy reality, especially when multiple institutions have investments in the material. A knot of competing copyright claims leave the digital collection in legal limbo: the musicians (or their estates) claim the rights to their performances; Smith's estate has claim to the reel-to-reel tapes, which live at the University of Arizona; while Duke University owns the digitized copies on the CDs. Streaming an audio archive for educational purposes would seem to fall under "fair use" in the United States, but the courts have interpreted this exception narrowly for audio recordings, and indeed what counts as an "educational purpose" is largely untested when it comes to sound. Moreover, the length of copyright protections—seventy years after the author's death—means that, realistically, much of the material in the Jazz Loft Project may not be available to the public for decades. And that's just the situation in the United States. It is often unclear what rights and responsibilities attend to an individual accessing U.S.-copyrighted materials for educational purposes from a physical location that is outside of the United States. Thus, outdated and ambiguous laws continually hamper the use of digital sounds in humanities research and teaching.

Even if the Jazz Loft Project were somehow able to overcome these seemingly insurmountable technological, institutional, and legal hurdles and could post the entire collection online for free public streaming, visitors would still face the challenge of finding discrete sounds and clips within four thousand hours of audio. Which is to say the collection is all but useless to researchers or even casual browsers without the textual metadata that Dan Partridge authored. Only through the intermediary of his knowledge and time—the hours he spent retuning his ears to the pitch of Smith's loft—did uninterpreted noise become keyword-searchable as the voice of Charles Mingus or a radio broadcast. Using pattern recognition to automate these search and discovery tasks in large corpora of audio is an active field of research, and it is possible that one day artificial intelligences will be able to take over for Dan, identifying Mingus's voice with minimal human intervention. For now, though, this labor is performed with human wetware, usually by a single cataloger (or a small group of catalogers) whose intellectual frameworks, interests, and knowledge of the subject shape the metadata and thus influence what type of research the collection supports. While digital media thus create a space of possibility for the study of sound, critical, interpretive labor fulfills this potential, not the technology itself.

As the Jazz Loft Project demonstrates, the humanities are in a moment of transition: between analog and digital; between the "old" methods and the "new"; between potential for change and the structures that hold it back. On the one hand, it has never been easier to build and access sonic archives or incorporate sound into scholarship. On the other, the ease with which sonic or audiovisual work can be shared and produced does not mean that academic writing, publishing, graduate training, or tenure and promotion have caught up with the possibilities. And so we—scholars of sound and technology—find ourselves at a crossroads. This book dwells in these various interstices as both a testament to the transformative value of experimenting with digital tools and a reinvestment in interpretive practices that always attend to the human. As our contributors demonstrate, amplifying the humanities through digital scholarship does not oppose close listening and deep historical analysis. Rather, these humanistic modes of interpretation provide the very foundation of digital sound studies.

The scratching and thumping that begin this essay perfectly encapsulate these tensions. When the cats batted Eugene Smith's microphone so many decades ago, the sounds that resonated in his loft were not the same as the commotion that we hear in the recording. Rather, they are "artifacts" of the technology itself: anomalies in the signal that draw our attention to the network of wires, transducers, and magnetic tape that enabled audio reproduction. By making audible the systems that are designed to be invisible—by letting us hear the presence of the microphone in the room—such glitches document the material conditions that make recording possible. The design of the microphone, its placement in Smith's loft, the nature of how those magnetic tapes encode and store sonic information, the altered nature of that information once it is digitized: these structures all shape sonic experience, whether we acknowledge them in our scholarship or not. This is true now more than ever, as digital technologies become both more ubiquitous and more entangled. Studying sound in the second decade of the twenty-first century demands that researchers pay critical attention to technologies, and especially to their invisibilities and silences.

No scholars are better placed to critically interpret, historicize, and seize the potential of the epistemological shifts brought about by the digital era than those who can interpret the cats' improvisational performance. The tools we use to listen to and reproduce sound are changing—along with forms of authorship and critical inquiry—and this book provides a blueprint for making sound central to research, teaching, and publishing practices. Using sound in one's work is not only imminently doable for humanities scholars today, it is, as this volume argues, urgently necessary. Digital sound studies holds the possibility of changing the text-centric and largely silent cultures of communication in the humanities into more richly multisensory experiences, inclusive of diverse knowledges and abilities.

Scholars have been carving out space for what we call digital sound studies for decades. Challenging the humanities to listen more closely—to attend, that is, not only to *what* but also to *how* we hear—sound studies scholars have productively theorized the sonic technologies that mediate and construct our experiences.[2] This growing body of research has taught us that sound has a politics; it can be gendered and racialized, used both to liberate people from and reinscribe determinative social categories. Sound has ethical implications and can help to build community or, conversely, to torment prisoners. It can elicit fear as easily as it produces longing or nostalgia. Even what counts as "sound" or "signal" and what gets dismissed as "noise" can differ dramatically across listening practices and auditory cultures.[3] Sound studies, then, places sounds in their cultural, historical, and social contexts. Dealing with the production, distribution, experience, poetics, or historicization of sound, as sound scholars have done, means dealing with the lived experiences of people.

One field has acknowledged the political complexity of sound since its inception: black studies. Generations of black cultural critics and authors have drawn deeply from music and sound in their writing. For instance, W. E. B. Du Bois frames each chapter in his classic *Souls of Black Folk* (1903) with excerpts from spirituals, which he theorizes as "sorrow songs" central to the African American experience.[4] Black studies has also had to confront sonically encoded racist stereotypes, such as those made popular in the United States through blackface minstrelsy and the use of "negro dialect" in early radio and television.[5] As a result scholars in the field have long been well attuned to the complex cultural significance of sound.[6] More recently, work at the intersection of sound studies and black studies has turned to technology to reveal its mediating effect on black aesthetic traditions. Fred Moten, for example, attends to the way the recording studio filters the philosophical conception of blackness in the work of Marvin Gaye.[7] Scholarship centered on popular music similarly assumes a form of culture making via technological reproduction, as can be seen in the work of Alexander Weheliye, Mark Anthony Neal, and Daphne Brooks.[8] In other cases, technology takes a more central role, as with Louise Meintjes's view of urban recording studios in South Africa that depicts the negotiation between races and ethnicities created by apartheid.[9] In this multidisciplinary body of work, scholars have shown that sound can serve many purposes: it can mobilize resistance, be a tactic of social negotiation, or contribute to structures of oppression and racialized representation.

The emergence of mechanical audio reproduction inspired scholars working within multiple fields to consider the social effects of mass distribution. This is especially true of cultural studies, where the technologization of sound was explored by many foundational theorists in the early to mid-twentieth century: Theodor Adorno, Walter Benjamin, and Roland

Barthes were followed by early media historians such as Walter Ong and Marshall McLuhan.[10] For these thinkers, the advent of new audiovisual technologies—the phonograph, film, radio, and eventually television—presented an opportunity to reconsider the relationship between technological and cultural production. Their work explores how the seemingly antihuman world of machines produces the modern political subject, extends the human body, and splits sounds from their sources, especially the human voice.[11] Some feared technology more than others. For instance, whereas Adorno (and later Jacques Attali) feared mass media's effect on culture, Benjamin seized on the power of the new medium of radio to disseminate ideas to the public, producing between eighty and ninety popular broadcasts on topics as wide-ranging as urban archaeology, literary tropes, and ancient history.[12] McLuhan, too, embraced popular media, making a cameo appearance in Woody Allen's *Annie Hall*. By treating audiovisual culture as a function of its technological reproduction, these early theorists laid the groundwork for the emergence of media and communication studies in the second half of the twentieth century.

A later generation of media scholars challenged the Marxian, modernist skepticism of technology that undergirds so much of this early work on the reproduction of sound. Technology is not non- or antihuman, they argued, but rather is always both producing and produced by human culture. That is, our listening practices are a product of the technologies that frame them, as much as the designs of our devices are shaped—literally—by the human body and the ways it listens.[13] Jonathan Sterne makes this point forcefully in *The Audible Past*, where he authors a cultural history of sound reproduction that upsets what he terms the "audiovisual litany"—the idea that sound and sight are mutually exclusive senses.[14] Other authors also explore the interconnectedness of sound, listening bodies, and technologies. Emily Thompson, writing about urban soundscape in the early twentieth century, reveals how mastery over sound in concert halls, churches, offices, and Hollywood soundstages was a cultural problem that sought technological solutions from the burgeoning field of acoustical science.[15] Lisa Gitelman attends to ways in which sound is always linked to multiple modes and media, showing the foundational role that visualist and tactile practices like reading, writing, and inscription played in the design of Edison's phonograph.[16] Together, this generation of media studies scholars reveals how the history of sound technology is always knit to the creation, production, and distribution of cultural memory and to the spaces of work, entertainment, and family.[17]

The wide-scale adoption of digital technologies at the end of the twentieth century brought a new set of concerns to the emerging field of sound studies, especially for those scholars who focus on music. Mark Katz and DJ Spooky, for instance, have situated seemingly "digital" practices like sampling within longer histories of sonic production, demonstrating the continuity between past and present.[18] Others, especially Tara Rodgers, have convincingly pushed for more inclusive histories of electronic music and the sound arts that include the contributions of women and people of color to the development of digital audio techniques.[19] Playback devices and instruments have been of particular interest, and Michael Bull's work on the iPod, Paul Théberge's work on synthesizers, and Mack Hagood's work on noise-canceling headphones elucidate how digital technologies mediate our relationship to sonic space in new ways.[20] Within and alongside research on digital music has flourished a renewed interested in materiality within media studies, especially the layered relationships between platforms, interfaces, and digital file formats.[21] Together, these digitally inflected approaches to sound ask media and digital studies scholars to think across software and hardware, and across forensic and formal materialities, and to continue to attend to the social and the cultural.

The fields of ethnomusicology, anthropology, and folklore also have their own long and storied relationship with technologies of sound. In the first half of the twentieth century, researchers in these nascent disciplines pioneered the use of portable recording equipment for collecting vernacular music.[22] The scripts they created for preserving sonic life influenced documentarians like Eugene Smith and survive today in the methods many ethnographers use to record their research in "the field." Early on, recording technology seemed to provide an efficient means to a noble end—preserving and venerating cultural forms that had previously been ignored. Over the years, however, it became clear that recording devices are not neutral mechanical objects: they play an agentive role in what is often a hierarchal encounter between researcher and subject. For instance, many prominent twentieth-century sound collectors were white scholars in positions of power making a living off of performances by rural, indigenous, and black and brown musicians.[23] In their recent returns to the early history of sound-based research, scholars Erika Brady, Benjamin Filene, Karl Hagstrom Miller, and others have illuminated the profoundly politicized nature of recording technologies as well as their lasting impact on the formation of academic fields, the music industry, and the preservation of vernacular culture in museums and archives.[24] Here, Steven Feld has been an innovator, composing soundscapes alongside more traditional print monographs to make explicit the way in which his own field recordings were always aesthetically manipulated.[25]

Because of the fraught histories of early sound collections, many of the institutions now housing them are grappling with how to preserve this material equitably in an era of mass digitization. Archivists and scholars —including Diane Thram, Sylvia Nannyonga-Tamusuza, Andrew N. Weintraub, and others—are asking what it might mean to repatriate digital sonic artifacts to their communities of origin.[26] Digitization would seem like a promising way to ensure that communities have access to their cultural heritage, but because reliable internet is a rare and costly commodity in many parts of the world, and especially in the global South, transmitting data online is untenable.[27] Furthermore, the history of economic exploitation surrounding much of this material means that some communities may not want their sonic artifacts to be widely available online. The U.S.-based Radio Haiti Archive is experimenting with disseminating digitized recordings from its collection to institutions and people in Haiti using USB sticks, a method of media transfer popular in areas where internet downloading and streaming are logistically difficult.[28] In an era when the vast majority of scholars are using digital devices on a regular basis, it is more important than ever to heed the lessons from our predecessors and carefully consider the ethical implications of seemingly benign technologies. For digital sound scholars, this means being particularly cognizant of the fact that internet access does not equate to universal access and being mindful that issues of power and publicity remain fraught.

As scholars of sound increasingly confront digital technologies, we find ourselves in conversation with digital humanities. Like sound studies, this interdisciplinary network encompasses a wide range of theories and practices loosely bound together by an interest in digital tools and technologies. On one end of its spectrum, critics such as Richard Grusin, Grant Wythoff, and others focus on culture and theory, drawing on methods from media and film studies to narrate the deep histories and philosophical implications of new technologies. Alex Galloway has clearly articulated the motivation behind such work in a recent interview with Melissa Dinsman: "The humanities needs to stop thinking of computation as an entirely foreign domain, and instead consider computers to be at the heart of what they have always done, that is, to understand society and culture as a technical and symbolic system."[29] Others within digital humanities take a more hands-on approach by building digital tools and platforms for humanities research. This work often emerges from lab-like research environments and includes projects such as Omeka, a curation platform for the web built at George Mason University's Roy Rosenzweig Center for History and New Media; Voyant Tools, a web-based text analysis platform built in collaboration between scholars at McGill University and the University of Alberta; and experimental text-visualization tools like Juxta and Ivanhoe, built at the University of Virginia's Institute for Advanced Technology in the Humanities. A particularly vibrant subfield of work right now, which can go by the name text mining or culturonomics, uses "big data" to analyze large bodies of text, image collections, and even audiovisual materials.[30]

At some moments these various strands of digital humanities have been antagonistic, and even the term "digital humanities" has created controversy. Some worry that the field has a far too comfortable relationship with systems of power that cultural criticism has long sought to challenge.[31] The scarcity of funding often exacerbates such tensions, especially in an era when the humanities are facing institutional pressure and falling enrollments. However, the digital turn has also reinvigorated conversations around the importance of humanities research and the often underappreciated, if not invisible, institutional structures that make our fields possible. For instance, digital humanities serves as a point of intersection between librarianship and scholarship, and libraries have become the de facto home for digital research on many campuses. These collaborations have led to the development of electronic collections that bring long-neglected authors and underrepresented histories to the public eye.[32] They have also galvanized discussions around the politics and long-term preservation of data in the humanities while advancing the cause of open access.[33] Publishing, too, has served as a point of intersection between different strands of work, as stakeholders across the humanities work together to develop digital platforms that speed up publication timelines and develop new protocols for peer review.[34] While the expansiveness of digital humanities, both as a field and as a "tactical" term that enables humanists to secure funding, has made it notoriously difficult to define, practitioners across all fields of study share an interest in exploring how digital media are transforming humanistic research.[35]

If sound studies and digital humanities have been confronting similar questions about praxis in the humanities and the nature of critical method, one might reasonably ask: Why has there been so little interest in sound within digital humanities? One answer lies in the text-centricity of the field, a bias that is baked into its institutional history. As a discipline, digital humanities locates its origin in Father Roberto Busa's *Index Thomisticus*, a concordance of every word in the works of Thomas Aquinas built using punch-card computing.[36] Its earliest journal is *Literary and Linguistic Computing*; among its earliest projects are electronic editions of literary works, leading to the formation of the Text Encoding Initiative (TEI) in the 1980s.[37] Digital humanities scholars generally communicate on Twitter and via long-running, heavily curated listservs like the Humanist rather than podcasts, favoring reading and typing over listening and speaking. While the early decades of the twenty-first century have seen the field expand significantly, including the creation of a new "AudioVisual Materials" Special Interest group of the Alliance for Digital Humanities Organizations, sound remains perhaps the least utilized, least studied mode within digital humanities. Few projects and fewer tools incite scholars to *listen*.

Yet this bias against sound is also a function of the nature of digital information itself. From the earliest days of personal computing, users interacted with machines through typed instructions issued through the command line. Vestiges of this interface are present in the ubiquitous search box of the web, where all content is parsed as a string of characters. Dependence on text within digital spaces persists in the user tagging that makes sound searchable on sites like SoundCloud and Genius, as well as in the more formal textual markup structures used to describe and organize digital content in projects like the Jazz Loft Archive. Simply put, making audio content accessible means rendering it as text. Even at its most abstract level, digital technology is built on a binary structure that mediates all data through strings of characters, which are then manipulated using text-based instructions. Thus even as we tend to imagine digital technologies as infinitely flexible, their fundamental unit is the discrete mark, the physical trace identified visually. This simple fact has given rise to a visualist orientation that continues to plague screen culture.

The silence of digital platforms has broader implications for teaching and research. Though rarely described as such, the sonic culture of the academy has always shaped what it is possible to know and to communicate. Many of the academy's most sacred practices involve the entanglement of text and oral performance, such as the dialogic and Socratic methods of lectures, conference papers, and colloquium presentations. Classrooms and seminars are inherently noisy spaces where students voice opinions, tap keyboards, and flip the pages of books. As much as focused study seems to be silent, the oral and the aural never recede from academic practices. Some digital platforms, like video conferencing, have amplified these aspects of academic communication, enabling scholars and students to speak across vast distances. Others silence our interactions. For example, many humanities scholars have criticized online learning for commodifying the education process, but collectively we must also recognize the impact of these changes on sensing bodies. Digital learning environments transform noisy spaces to silent screens, where students interact with their instructors and classmates almost entirely via written language. The proliferation of silence via text-oriented digital technologies affects individual learners and educators differently.

What forms of knowledge—and what embodied experiences—are diminished by the humanities' reliance on text and visualist methods? And whose voices are going unheard in the digital turn? Bringing sound studies into meaningful conversation with digital humanities has the power to inspire new questions and foment new methods that are radically different from those of print. By foregrounding sonic experience, this collection begins answering these questions, using auditory culture to probe the assumptions of digital tools and technologies in academic life. Engaging deeply with sound, as our contributors collectively argue, untethers scholars from their reliance on text-based modes of knowledge, revealing the structural biases built into the apparatus of scholarship and transforming the epistemic grounds upon which such conversations can be had.

Publishing venues and researchers are already challenging the biases of the contemporary media environment through multimodal scholarship. A variety of journals including *Kairos*, *Liminalities*, and *Computers and Composition Online*, blogs like *Sounding Out!*, and platforms such as Scalar have created venues for born-digital work that encourage exploration and experimentation while building on established traditions of academic writing and argumentation.[38] The creative use of new media is at play in a number of projects that combine audio with a wide range of digitally archived material. Sharon Daniel's "Public Secrets" is an interactive (and intentionally public) audio archive of interviews with incarcerated women who pointedly describe the prison industrial complex and its injustices.[39] The historically focused Freedom's Ring, a product of the Martin Luther King, Jr., Research and Education Institute at Stanford University, mirrors the audio from King's iconic "I Have a Dream" speech with his written draft so that users experience both versions of the speech simultaneously. An "index" links this audiovisual rendering of King's speech to a number of digitized archival documents relevant to the performance and its political moment.[40] Similarly, Emily Thompson's "The Roaring Twenties," a complement to her monograph *The Soundscape of Modernity*, employs New York City noise ordinances in the 1920s to explore everyday contestations of the urban soundscape.[41] These innovative projects create reading and listening experiences that give agency to the user, thereby challenging the unidirectionality of conventional scholarly writing. It is also significant that each project was created collaboratively: Freedom's Ring was developed under the direction of Evan Bissell in partnership with Beacon Press's King Legacy Series and the MLK Institute at Stanford; Thompson's with the help of web designer Scott Mahoney; and Daniel's with support from the design team at *Vectors* journal. Like much digital humanities work, digital sound studies is changing the model for academic production by moving away from single-authored, single-argument work toward collaborative, multimodal projects that allow for multiple pathways and target broad audiences.

This volume cuts across the wide-ranging disciplines engaged in sound-based research, encompassing literature, performance, disability, anthropology, black studies, history, information science, and more. However, the contributors refrain from engaging solely in field-specific debates, speaking instead to the broader issues, opportunities, and challenges that emerge from thinking about and with sound in digital environments. Part 1, "Theories and Genealogies," lays the historical and conceptual groundwork for this exploration by linking digital sound studies to important shifts in academic thought and practice that took place in the twentieth century. Historian Richard Cullen Rath narrates the history of his encounters with digital methods, beginning with his experiences as a student. For more than two decades he has studied a rare historical document of African-diasporic music in Jamaica. An early adopter of MIDI technology, over the years Rath has combined digital and analog methods to create playable historical replicas of instruments and to interpret the music. This essay meditates on the importance of digitally informed "ethnohistory" for illuminating the cultural contributions of enslaved Africans and subaltern histories.

Myron Beasley anchors digital sound studies praxis in the critical moves of black radicalism and embodied performance. In an engaging narrative that unfolds like tracks on an album, Beasley draws on Zora Neale Hurston's work to show how her innovative uses of technology to record folk culture in her native Florida connect to the performance of a DJ sampling her voice on a laptop in a Harlem cafe. Beasley also explores the politics of metadata and the problems caused by the way archives misrepresent Hurston's scholarship by identifying her work with her white male colleagues. This chapter thus narrates a genealogy of digital sound studies rooted in black feminist theory, performance, and ethnographic practice. Through an exploration of Walter Ong's theories of orality and rhetoric, Jon Stone's essay also explores how sound often operates as the connective tissue at this particular moment of technological hybridity when the term "digital" signals work that is participatory, spontaneous, and often noisy. The essay begins with Stone's encounter with a single digital sound object: a YouTube video of CHOIR! CHOIR! CHOIR! (an ad hoc vocal ensemble in Toronto) performing Phil Collins's "In the Air Tonight." Stone riffs on the digitally mediated performance to introduce what he calls "digital humanity"—the connective potential of today's technologies.

Stone's essay delivers readers to part 2, which highlights the way scholars are using social media and digital pedagogy to build communities of thought around sonic research. The editorial team behind the *Sounding Out!* blog single-handedly transformed the look, feel, and sound of contemporary sound studies by instigating a conversation online that unites a wide-ranging field. Importantly, they have brought voices from the margins into the center by curating and promoting sound studies work through the site's social media presence. In their essay, Aaron Trammell, Jennifer Stoever, and Liana Silva examine the affective labor entailed in the act of building a strong digital community and provide a biography of their project. Regina Bradley's series of YouTube interviews about the significance of the music group OutKast similarly shifts the conversation in her field to be more inclusive of regionalisms of the American South in the study of hip-hop. She reached new public and academic audiences while building a multimedia archive of cultural criticism. In her essay, she documents the intellectual outcomes of this work and creates a template for others wishing to embark on a similar method of digital sound research and publication. W. F. Umi Hsu brings this ethos of community building to the classroom, where they ask their students to engage in audio-ethnography in collaboration with local middle schoolers. By producing sound recordings in collaboration with community partners, Hsu's students learn that sonic methods can challenge hierarchies and build bridges across cultures and generations. Hsu explores their students' insights and experiences to demonstrate that turning to sound amplifies the already transformational aspects of digital pedagogy.

Each of the scholars in part 3, "Disciplinary Translations," traverses boundaries to build new conceptual frameworks for digital sound studies. In her essay, Tanya Clement explains that the metadata conventions of information science create significant barriers for data-driven digital sound scholarship. Clement is the principal investigator of the NEH-funded project High Performance Sound Technologies for Access and Scholarship (HiPSTAS), which aims to harness the capacity of big-data analytics for the study of spoken-word audio collections. Clement's investigation is crucial for securing the potential for digital sound studies to enhance the research potential of large audio collections through innovative computational analysis and discovery. Yet her observations remain rooted in practices of close listening that attend to the nuances of sonic meaning in cultural life.

Michael Kramer takes aim at the frustrating ubiquity of visualization techniques in digital humanities by flipping the script and remediating visual media such as maps and photographs as sonic data. His avant-garde methods of "sonification" demonstrate that sound-based research can be meaningful for scholars working with visual culture. A historian by training, Kramer listens to seemingly "silent" visual artifacts from the historic Berkeley Folk Festival archive, showing how to interpret the sounds encoded in images through a deeply multimodal praxis. Trained in literary studies, and a researcher of Victorian music, Joanna Swafford shows how digital methods enable her to present her work to different disciplines. Faced with the challenges of writing about the nuances of musical notation for a literary audience, she designed an open-source tool, Augmented Notes, that makes it possible for people who do not read music to learn more about the relationship between musical scores and performance. Her digital solution, however, has multiple potential applications that may be used across fields to animate notational music for a variety of purposes.

Part 4 "Points Forward" to the next wave of digital sound scholarship by identifying key challenges that the field needs to address. Digital humanists are just beginning to develop methods of assessment and evaluation that recreate the rigor of peer review, a practice not without its own critics.[42] Rebecca Geoffroy-Schwinden identifies what makes digital scholarship about historical sounds effective while reviewing key projects that examine the cultural history of sound. She also narrates her own efforts to bring to life the music of the French Revolution on the platform Scalar. Geoffroy-Schwinden argues that digital explorations of sonic history must do more than simply attempt to recreate the sounds of the past; these projects must also contextualize the listening perspectives of historical subjects. She shows that without understanding what made sound interpretable and meaningful to those who produced and heard it, even cutting-edge digital work fails to live up to its promise. Finally, Steph Ceraso considers the multisensory aspects of sonic experience as a means of rethinking ways to incorporate sound into born-digital scholarship. Beginning with observations from her own work, she offers three "sound practices" for helping scholars recognize the multifaceted ways in which sound is embodied. She tackles a range of issues—from universal design to the tactility of sound—as a means of illustrating a simple but powerful point: the work of digital sound studies necessitates creative thinking that pushes against conventional wisdom.

In an afterword on the futures of digital sound studies, Jonathan Sterne responds to the collection in an interview with the editors. This conversation —a print remediation of a Skype session that occurred in four different places at once—reflects on the shifts in both academic and technological culture that brought us to this moment. Sterne discusses the institutional infrastructures that will need to change in order to sustain the momentum behind work at the juncture of sound studies and digital humanities. He also identifies themes humming behind each of the essays in terms of digital publication—the platforms that enable it and its relationship to academic prestige. This interview, as with the rest of this volume, is a textual artifact of digital sonic practice.

Sterne's commentary is a fitting place to end as it broadens the conversation to examine the institutional frameworks that make digital sound studies possible. For multimodal scholarship to continue to grow, it must be met with significant institutional imagination and collaboration. Scholars need librarians to aid with accessing and archiving digital materials to ensure the long-term preservation and sustainability of emerging forms of scholarship. Librarians need the financial and organizational support of their universities, and they need an open line of communication with academic publishers and for-profit companies about the possibilities and limitations of electronic scholarship. Administrators need to be shown, and to recognize when shown, the intellectual value of formal experimentation and creativity within the broader goals of the humanities. Mentors need to encourage junior scholars to take risks while clearly apprising them of what they stand to gain, as well as what they may lose, within their particular institutional cultures and career trajectories. Educators need training, time, and professional development to begin learning how to integrate new technologies into the classroom in ways that prepare students to be active participants in twenty-first-century media cultures without losing sight of the core values of the humanities. Navigating this dense network of stakeholders is difficult and often risky work, especially for junior scholars who increasingly find themselves needing to abandon the advice of senior academics and forge a path for their own future within a rapidly changing discipline.

The contributors in this volume are doing just that. By being students of their own cultural moment, they harness the transformative potential of digital technologies and platforms to amplify underrepresented voices, write alternative histories, reimagine the classroom experience, and design capacious new modes of scholarship and publishing. That is to say, digital sound studies scholars combine the creative use of sonic technologies with an informed critical inquiry of them, merging the lessons of digital humanities and the "maker" movement with a thoughtful analysis of digital culture, new media, and the sonic possibilities of technologized learning spaces.[43]

Sonic technologies are not unified objects with clear intent or singular uses; rather, they are always open to appropriation by users whose actions transform the technology itself. Just as the portable reel-to-reel recorder catalyzed Eugene Smith's project, the proliferation of digital technologies creates a space for sound scholars to revisit the media and modes that motivate all stages of the research process. Digital sound scholars are tinkerers, inventors, explorers, and collaborators whose experimentations with new forms of knowledge production transform diverse fields while transcending disciplinary borders. As sound scholars draw on the innovations of digital humanities and, in turn, digital humanities becomes amplified, digital sound studies enriches the academy as a whole with the power of sonic experience.

NOTES

1 After discovering Smith's tapes at the University of Arizona, the Jazz Loft Project's program director, Sam Stephenson, spearheaded efforts to preserve them at Duke University's Center for Documentary Studies, where we listened to the newly digitized reel-to-reel recordings. The digital collection is now housed at Duke's Rubenstein Rare Book and Manuscript Library. For more on the history of the collection, see Stephenson, *Jazz Loft Project*. Some audio recordings can be heard on the project's website, www.jazzloftproject.org (accessed January 13, 2018).

2 Since 2003, several key sound studies volumes and collections have been published. For a view of the field's history, see the introduction in Sterne, *Sound*

Studies Reader. Back and Bull's Auditory Culture Reader takes a cultural studies approach, while Pinch and Bijsterveld's Oxford Handbook of Sound Studies focuses more on media and technology. Erlmann's Hearing Cultures and Smith's Hearing History gather historical work on sound; Novak and Sakakeeny's Keywords in Sound emphasizes ethnography. For perspectives from film studies, a significant precursor to the emergence of sound studies, see Beck and Grajeda's Lowering the Boom. The largest and most comprehensive edited volume that covers many overlapping subjects—for example, culture, ecology, listening, sound and space, and media (television, film, radio)—is the four-volume set Sound Studies, also edited by Michael Bull. We are indebted to Brian Kane, on the sound studies listserv on Google Groups, who suggested that differentiating sound studies anthologies according to their scholarly perspectives would be helpful.

3 For more about the politics of noise, see Attali, Noise; Goodman, Sonic Warfare; Cusick, "An Acoustemology of Detention"; Novak, Japanoise; Cusick, "'You are in a place that is out of the world.'"

4 Du Bois, Souls of Black Folk.

5 On the cultural legacy of blackface minstrelsy, see Lott, Love and Theft, and Lhamon, Raising Cain. On the way dialect affected major spoken-word audio collections, see Taylor, "Saving Sound, Sounding Black."

6 Much of this this work examines the intersection of music and culture, albeit with a Euro-American bias. See Douglass, Narrative of the Life; Baraka, Blues People; Ellison and O'Meally, Living with Music; Davis, Blues Legacies and Black Feminism; and Southern, Music of Black Americans. The weight toward North America and Europe of this work is indicative of sound studies as a whole. Some exceptions, mostly from ethnomusicology, are Feld, Sound and Sentiment; Meintjes, Sound of Africa!; Novak, Japanoise; and Ochoa, Aurality. More recently, Gustavus Stadler criticized mainstream sound studies scholarship for having a significant race problem deriving from its own associations with technoculture; see Stadler, "On Whiteness and Sound Studies."

7 Moten, In the Break, 171–232.

8 Weheliye, Phonographies; Neal, What the Music Said, Soul Babies, and Songs in the Key of Black Life; and Brooks, "Nina Simone's Triple Play."

9 Meintjes, Sound of Africa!

10 Benjamin, "Work of Art in the Age of Mechanical Reproduction"; Adorno, Essays on Music; Barthes, Image, Music, Text; Ong, Orality and Literacy; and McLuhan, Understanding Media.

11 An in-depth exploration of this question can be found in Attali, Noise, and Jameson, "Postmodernism and Consumer Society." See also Deleuze, Difference and Repetition, and Chion, Audio-Vision.

12 Many transcriptions of these shows can be found in Benjamin and Rosenthal, Radio Benjamin.

13 For more on the body as a "sensing agent," see Helmreich, *Sounding the Limits*; Eidsheim, *Sensing Sound*; and Erlmann, "But What of the Ethnographic Ear?"
14 Sterne, *Audible Past*, 15–16.
15 Thompson, *Soundscape of Modernity*.
16 Gitelman, *Scripts, Grooves*.
17 For more on sonic technologies and the cultural practices of remembering, see Bijsterveld, *Sound Souvenirs*.
18 Katz, *Capturing Sound*, and Paul Miller, *Sound Unbound*.
19 Rodgers, *Pink Noises*.
20 Bull, *Sound Moves*; Théberge, *Any Sound You Can Imagine*; and Hagood, "Quiet Comfort." For more on the synthesizer, also see Evens, *Sound Ideas*.
21 For an introduction to such work, see the companion website to the Platform Studies series by MIT Press, edited by Ian Bogost and Nick Montfort (accessed January 13, 2018, http://platformstudies.com). For a thorough discussion of platform theory as it relates to audio technologies and cultures, see Sterne, *MP3*.
22 Portable recording devices have played a significant role in ornithology, too, enabling scientists and sound archivists, such as those at the Macaulay Library, to build large research collections of animal sounds recorded around the world by both experts and amateur bird enthusiasts. For more on the history of nature recordings, see Bruyninckx, "Sound Sterile," and Eley, "A Birdlike Act."
23 The long history of representing performance traditions of indigenes, underclasses, and colonial others emerged from travel writing of the colonial period and assumes a distinct character with the rise of blackface minstrelsy in the nineteenth century. For more on the recording of African diasporic music in musical notation by white authors, see Radano, *Lying Up a Nation*, 164–229. For a digital sound project on one of these early works, see Dubois, Garner, and Lingold, *Musical Passage*.
24 Brady, *A Spiral Way*; Filene, *Romancing the Folk*; and Karl Miller, *Segregating Sound*. Scholars working on performance traditions from before the dawn of sound recording know all too well the constraints that technologies impose on research possibilities. Applied ethnomusicologists approach this problem by maintaining musical ensembles of traditional music, using performance as a form of public archive. See Harrison and Pettan, *Applied Ethnomusicology*; Harrison, "Epistemologies of Applied Ethnomusicology"; and Seeger, "Lost Lineages."
25 One example is Feld, *Voices of the Rainforest*, which is discussed in Feld, "A Sweet Lullaby." Scholars and libraries operating in the public sphere also have explored alternative ways of presenting sound. These include R. Murray Schafer's World Soundscape Project (an acoustic ecology project founded in the late 1960s) and the more recent activities at the Library of Congress's American Memory Project and the British Library Sound Archives. For more

on soundscapes, see Schafer, *Tuning of the World* and *Book of Noise*; Truax, *World Soundscape Project's Handbook*; and Harley, Minevich, and Waterman, *Art of Immersive Soundscapes*. Thanks to Steph Ceraso and Jonathan Sterne for pointing us toward these resources.

26 For some perspectives regarding these challenges, see Nannyonga-Tamusuza and Weintraub, "The Audible Future"; Thram, "Performing the Archive"; and Nannyonga-Tamusuza, "Documentation of the Wachsmann Collection."

27 For one investigation into the circulation of digital music in the global South, see Steingo, "Sound and Circulation."

28 Wagner, "Bringing Radio Haiti Home."

29 Dinsman and Galloway, "Digital in the Humanities."

30 On quantitative analysis in literary studies, see Moretti, *Graphs, Maps, Trees*; Jockers, *Macroanalysis*; and Underwood, *Why Literary Periods Mattered*, and his blog, the Stone and the Shell. Several university collectives are currently exploring data analysis approaches to the history of literature, including the Chicago Text Lab, the Stanford Literary Lab, and the .txtLAB at McGill. For interdisciplinary perspectives on distant reading in art and sound studies, respectively, see Manovich, "How to Compare One Million Images?," and Clement, "Distant Listening," as well as Clement's and Kramer's essays in this volume.

31 See, for instance, the work of the #transformDH collective and essays in the special issue of *differences* regarding the "Dark Side of Digital Humanities," especially McPherson, "Designing for Difference," and Barnett, "Brave Side of Digital Humanities."

32 Hartsell-Grundy, Braunstein, and Golomb, *Digital Humanities in the Library*.

33 Klein, *Interdisciplining Digital Humanities*.

34 For example, see Humanities Commons, a web-based networking platform for humanities scholars to share their research (accessed January 13, 2018, https://hcommons.org). On digital humanities and peer review, see Fitzpatrick, *Planned Obsolescence*.

35 On "tactical" digital humanities, see Kirschenbaum, "Digital Humanities As/Is." For a more general introduction to the field and its debates, see Gold, *Debates in the Digital Humanities*; Berry, *Understanding Digital Humanities*; Jones, *Emergence of Digital Humanities*; and Svensson and Goldberg, *Between Humanities and the Digital*.

36 See Jones, *Emergence of Digital Humanities*.

37 On the history of digital humanities, see Hockey, "History of Humanities Computing."

38 Scalar was created by the Alliance for Networking Visual Culture. Born out of a desire to integrate film excerpts more seamlessly into academic writing, the platform boasts sophisticated tools for including audio and visual material within digital texts. For more information, see their website, http://scalar.usc.edu/scalar (accessed January 13, 2018). Several other academic outlets, includ-

ing *Harlot*, have experimented with multimodal scholarship. For example, see Ceraso and Stone, "Sonic Rhetorics," *Harlot*'s special issue on sound.
39 Daniel, "Public Secrets."
40 Bissell, *Freedom's Ring*.
41 Other recent examples of web projects featuring sound include the London Sound Survey; McDonald, Every Noise at Once; and Wall, Virtual Paul's Cross Project.
42 Fitzpatrick, *Planned Obsolescence*.
43 For more about maker culture, see Ratto and Boler, DIY *Citizenship*, and Ratto, "Critical Making." Also see the accompanying website to the Maker Lab in the Humanities (MLab) at the University of Victoria, directed by Jentery Sayers (accessed January 13, 2018, http://maker.uvic.ca).

WORKS CITED

Adorno, Theodor W. *Essays on Music*. Introduction, commentary, and notes by Richard D. Leppert. Berkeley: University of California Press, 2002.

Attali, Jacques. *Noise: The Political Economy of Music*. Minneapolis: University of Minnesota Press, 1985.

Back, Les, and Michael Bull, eds. *The Auditory Culture Reader*. New York: Berg, 2003.

Baraka, Amiri. *Blues People: Negro Music in White America*. New York: W. Morrow, 1963.

Barnett, Fiona M. "The Brave Side of Digital Humanities." Special issue, "Dark Side of the Digital Humanities," *differences: A Journal of Feminist Cultural Studies* 25, no. 1 (2014): 64–78.

Barthes, Roland. *Image, Music, Text*. New York: Noonday Press, 1977.

Beck, Jay, and Tony Grajeda, eds. *Lowering the Boom: Critical Studies in Film Sound*. Urbana: University of Illinois Press, 2008.

Benjamin, Walter. "The Work of Art in the Age of Mechanical Reproduction." In *Illuminations*, edited by Hannah Arendt. New York: Schocken Books, 1986.

Benjamin, Walter, and Lecia Rosenthal. *Radio Benjamin*. New York: Verso, 2014.

Berry, David, ed. *Understanding Digital Humanities*. New York: Palgrave Macmillan, 2012.

Bijsterveld, Karin, ed. *Sound Souvenirs: Audio Technologies, Memory, and Cultural Practices*. Amsterdam: Amsterdam University Press, 2009.

Bissell, Evan, dir. Freedom's Ring. Accessed June 10, 2014. http://freedoms-ring.org/?view=Speech.

Brady, Erika. *A Spiral Way: How the Phonograph Changed Ethnography*. Jackson: University Press of Mississippi, 1999.

Brooks, Daphne A. "Nina Simone's Triple Play." *Callaloo* 34, no. 1 (2011): 176–97.

Bruyninckx, Joeri. "Sound Sterile: Making Scientific Field Recordings in Ornithology." In *The Oxford Handbook of Sound Studies*, edited by Trevor Pinch and Karin Bijsterveld. New York: Oxford University Press, 2012.

Bull, Michael. *Sound Moves: iPod Culture and Urban Experience*. New York: Routledge, 2007.

Bull, Michael, ed. *Sound Studies: Critical Concepts in Media and Cultural Studies*. New York: Routledge, 2013.

Ceraso, Steph, and Jon Stone, eds. "Sonic Rhetorics." Special issue, *Harlot* 9 (2013). Accessed June 10, 2014. http://harlotofthearts.org/index.php/harlot/issue/view/9.

Chion, Michel. *Audio-Vision: Sound on Screen*. Translated by Claudia Gorbman. New York: Columbia University Press, 1994.

Clement, Tanya, et al. "Distant Listening to Gertrude Stein's 'Melanctha': Using Similarity Analysis in a Discovery Paradigm to Analyze Prosody and Author Influence." *LLC: Journal of Digital Scholarship in the Humanities* 28, no.4 (2013): 582–602.

Cusick, Suzanne. "'You are in a place that is out of the world': Music in the Detention Camps of the 'Global War on Terror.'" *Journal of the Society for American Music* 2, no. 1 (2008): 1–26.

Cusick, Suzanne. "An Acoustemology of Detention in the 'Global War on Terror.'" In *Music, Sound and the Reconfiguration of Public and Private Space*, edited by Georgina Born. New York: Cambridge University Press, 2013.

Daniel, Sharon. "Public Secrets." *Vectors: Journal of Culture and Technology in a Dynamic Vernacular*, Perception Issue, 2, no. 2 (2007): n.p. http://vectors.usc.edu/projects/index.php?project=57.

Davis, Angela Y. *Blues Legacies and Black Feminism: Gertrude "Ma" Rainey, Bessie Smith, and Billie Holiday*. New York: Pantheon Books, 1998.

Deleuze, Gilles. *Difference and Repetition*. New York: Columbia University Press, 1994.

Dinsman, Melissa, and Alexander R. Galloway. "The Digital in the Humanities: An Interview with Alexander Galloway." *Los Angeles Review of Books*, March 27, 2016. https://lareviewofbooks.org/article/the-digital-in-the-humanities-an-interview-with-alexander-galloway.

Douglass, Frederick. *Narrative of the Life of Frederick Douglass: An American Slave*. Boston: Published at the Anti-Slavery Office, 1845.

Dubois, Laurent, David K. Garner, and Mary Caton Lingold. *Musical Passage: A Voyage to 1688 Jamaica*. Accessed September 5, 2016. http://musicalpassage.org.

Du Bois, W. E. B. *The Souls of Black Folk*. New York: Dover, 1994.

Eidsheim, Nina. *Sensing Sound: Singing and Listening as Vibrational Practice*. Durham, NC: Duke University Press, 2015.

Eley, Craig. "A Birdlike Act: Sound Recording, Nature Imitation, and Performance Whistling." *Velvet Light Trap* 74 (Fall 2014): 4–15.

Ellison, Ralph, and Robert G. O'Meally. *Living with Music*. New York: Random House, 2002.

Erlmann, Veit, ed. "But What of the Ethnographic Ear? Anthropology, Sound, and the Senses." In *Hearing Cultures: Essays on Sound, Listening, and Modernity*, edited by Veit Erlmann, 1–20. New York: Berg, 2004.

Evens, Aden. *Sound Ideas: Music, Machines, and Experience*. Minneapolis: University of Minnesota Press, 2005.

Feld, Steven. *Sound and Sentiment: Birds, Weeping, Poetics, and Song in Kaluli Expression*. Philadelphia: University of Pennsylvania Press, 1982.

Feld, Steven. "A Sweet Lullaby for World Music." *Public Culture* 12, no. 1 (2000): 145–71.

Feld, Steven. *Voices of the Rainforest*. Rykodisc, RCD 10173, 1991.

Filene, Benjamin. *Romancing the Folk: Public Memory and American Roots Music*. Chapel Hill: University of North Carolina Press, 2000.

Fitzpatrick, Kathleen. *Planned Obsolescence: Publishing, Technology, and the Future of the Academy*. New York: New York University Press, 2011.

Gitelman, Lisa. *Scripts, Grooves, and Writing Machines: Representing Technology in the Edison Era*. Palo Alto, CA: Stanford University Press, 1999.

Gold, Matthew K. *Debates in the Digital Humanities*. Minneapolis: University of Minnesota Press, 2012.

Goodman, Steve. *Sonic Warfare: Sound, Affect, and the Ecology of Fear*. Cambridge, MA: MIT Press, 2010.

Hagood, Mack. "Quiet Comfort: Noise, Otherness, and the Mobile Production of Personal Space." *American Quarterly* 63, no. 3 (2011): 573–89.

Harley, James, Pauline Minevich, and Ellen Waterman, eds. *Art of Immersive Soundscapes*. Regina, Saskatchewan: University of Regina Press, 2013.

Harrison, Klisala. "Epistemologies of Applied Ethnomusicology." *Ethnomusicology* 56, no. 3 (2012): 505–29.

Harrison, Klisala, and Svanibor Pettan. *Applied Ethnomusicology: Historical and Contemporary Approaches*. Newcastle-Upon-Tyne, UK: Cambridge Scholars Publishing, 2010.

Hartsell-Grundy, Arianne, Laura Braunstein, and Liorah Golomb, eds. *Digital Humanities in the Library: Challenges and Opportunities for Subject Specialists*. Chicago: Association of College and Research Libraries, 2015.

Helmreich, Stefan. *Sounding the Limits of Life: Essays in the Anthropology of Biology and Beyond*. Princeton, NJ: Princeton University Press, 2016.

Hockey, Susan. "The History of Humanities Computing." In *A Companion to Digital Humanities*, edited by Susan Schreibman, Ray Siemens, and John Unsworth. Oxford, UK: Blackwell, 2004.

Jacqueline, Wernimont. "Hearing Eugenics." *Sounding Out!*, July 18, 2016. https://soundstudiesblog.com/2016/07/18/hearing-eugenics.

Jameson, Fredric. "Postmodernism and Consumer Society." In *Postmodernism and Its Discontents*, edited by E. Ann Kaplan. New York: Verso, 1988.

Jockers, Matthew. *Macroanalysis: Digital Methods and Literary History*. Urbana: University of Illinois Press, 2013.

Jones, Steven E. *The Emergence of Digital Humanities*. New York: Routledge, 2014.

Katz, Mark. *Capturing Sound: How Technology Has Changed Music*. Berkeley: University of California Press, 2004.

Kirschenbaum, Matthew. "Digital Humanities As/Is a Tactical Term." In *Debates in the Digital Humanities*, edited by Matthew K. Gold. Minneapolis: University of Minnesota Press, 2012.

Klein, Julie Thompson. *Interdisciplining Digital Humanities: Boundary Work in an Emerging Field*. Ann Arbor: University of Michigan Press, 2015.

Lhamon, W. T. *Raising Cain: Blackface Performance from Jim Crow to Hip Hop*. Cambridge, MA: Harvard University Press, 1998.

Lott, Eric. *Love and Theft: Blackface Minstrelsy and the American Working Class*. New York: Oxford University Press, 1995.

Manovich, Lev. "How to Compare One Million Images?" In *Understanding Digital Humanities*, edited by David Berry. New York: Palgrave Macmillan, 2012.

McDonald, Glenn. Every Noise at Once. Accessed July 6, 2015. http://everynoise.com/engenremap.html.

McPherson, Tara. "Designing for Difference." Special issue, "Dark Side of the Digital Humanities." *differences: A Journal of Feminist Cultural Studies* 25, no. 1 (2014): 177–88.

McLuhan, Marshall. *Understanding Media: The Extensions of Man*. Cambridge, MA: MIT Press, 1994.

Meintjes, Louise. *Sound of Africa! Making Music Zulu in a South African Studio*. Durham, NC: Duke University Press, 2003.

Miller, Karl Hagstrom. *Segregating Sound: Inventing Folk and Pop Music in the Age of Jim Crow*. Durham, NC: Duke University Press, 2010.

Miller, Paul (a.k.a. DJ Spooky That Subliminal Kid). *Sound Unbound: Sampling Digital Music and Culture*. Cambridge, MA: MIT Press, 2008.

Moretti, Franco. *Graphs, Maps, Trees: Abstract Models for Literary History*. New York: Verso, 2005.

Moten, Fred. *In the Break: The Aesthetics of the Black Radical Tradition*. Minneapolis: University of Minnesota Press, 2003.

Nannyonga-Tamusuza, Sylvia. "Written Documentation of the Klaus Wachsmann Music Collection: Repatriating the Past to Present Indigenous Users in Uganda." In *African Musics in Context: Institutions, Culture, Identity*, edited by Thomas Solomon. Kampala, Uganda: Fountain Publishers, 2015.

Nannyonga-Tamusuza, Sylvia, and Andrew N. Weintraub. "The Audible Future: Reimagining the Role of Sound Archives and Sound Repatriation in Uganda." *Ethnomusicology* 56, no. 2 (Spring 2012): 206–33.

Neal, Mark Anthony. *Songs in the Key of Black Life: A Rhythm and Blues Nation*. New York: Routledge, 2003.

Neal, Mark Anthony. *Soul Babies: Black Popular Culture and the Post-Soul Aesthetic*. New York: Routledge, 2002.

Neal, Mark Anthony. *What the Music Said: Black Popular Music and Black Public Culture*. New York: Routledge, 1999.

Novak, David. *Japanoise: Music at the Edge of Circulation*. Durham, NC: Duke University Press, 2013.

Novak, David, and Matt Sakakeeny, eds. *Keywords in Sound*. Durham, NC: Duke University Press, 2015.

Ochoa Gautier, Ana María. *Aurality: Listening and Knowledge in Nineteenth-Century Colombia*. Durham, NC: Duke University Press, 2014.

Ong, Walter J. *Orality and Literacy: The Technologizing of the Word*. New York: Methuen, 1982.

Pinch, Trevor, and Karin Bijsterveld, eds. *The Oxford Handbook of Sound Studies*. New York: Oxford University Press, 2012.

Radano, Ronald M. *Lying Up a Nation: Race and Black Music*. Chicago: University of Chicago Press, 2003.

Ratto, Matt, and Megan Boler, eds. *DIY Citizenship: Critical Making and Social Media*. Cambridge, MA: MIT Press, 2014.

Ratto, Matt. "Critical Making: Conceptual and Material Studies in Technology and Social Life." *Information Society* 27, no. 4 (2011): n.p. https://doi.org/10.1080/01972243.2011.583819.

Rawes, Ian. The London Sound Survey. Accessed July 6, 2015. www.soundsurvey.org.uk.

Rodgers, Tara. *Pink Noises: Women on Electronic Music and Sound*. Durham, NC: Duke University Press, 2010.

Schafer, R. Murray. *The Book of Noise*. Indian River, Ontario: Commercial Press, [1968] 1998.

Schafer, R. Murray. *The Tuning of the World*. New York: Knopf, 1977.

Seeger, Anthony. "Lost Lineages and Neglected Peers: Ethnomusicologists Outside Academia." *Ethnomusicology* 50, no. 2 (2006): 214–35.

Smith, Mark M., ed. *Hearing History: A Reader*. Athens: University of Georgia Press, 2004.

Southern, Eileen. *The Music of Black Americans: A History*. New York: Norton, 1971.

Stadler, Gustavus. "On Whiteness and Sound Studies." *Sounding Out!*, July 6, 2015. http://soundstudiesblog.com/2015/07/06/on-whiteness-and-sound-studies.

Steingo, Gavin. "Sound and Circulation: Immobility and Obduracy in South African Electronic Music." *Ethnomusicology Forum* 24, no. 1 (2015): 102–23.

Stephenson, Sam. *The Jazz Loft Project*. Durham, NC: Duke University Press, 2009.

Sterne, Jonathan. *The Audible Past: Cultural Origins of Sound Reproduction*. Durham, NC: Duke University Press, 2003.

Sterne, Jonathan. *MP3: The Meaning of a Format*. Durham, NC: Duke University Press, 2012.

Sterne, Jonathan, ed. *The Sound Studies Reader*. New York: Routledge, 2012.

Svensson, Patrik. "Beyond the Big Tent." In *Debates in the Digital Humanities*, edited by Matthew K. Gold. Minneapolis: University of Minnesota Press, 2012.

Svensson, Patrik, and David Theo Goldberg. *Between Humanities and the Digital*. Cambridge, MA: MIT Press, 2015.

Taylor, Toniesha L. "Saving Sound, Sounding Black, Voicing America: John Lomax and the Creation of the 'American Voice.'" *Sounding Out!*, June 8, 2015. http://soundstudiesblog.com/2015/06/08/john-lomax-and-the-creation-of-the-american-voice.

Théberge, Paul. *Any Sound You Can Imagine: Making Music/Consuming Technology*. Hanover, NH: Wesleyan University Press, 1997.

Thompson, Emily Ann. *The Soundscape of Modernity: Architectural Acoustics and the Culture of Listening in America, 1900–1933*. Cambridge, MA: MIT Press 2002.

Thram, Diane. "Performing the Archive: Repatriation of Digital Heritage and the ILAM Music Heritage Project SA." In *African Musics in Context: Institutions, Culture, Identity*, edited by Thomas Solomon. Kampala, Uganda: Fountain Publishers, 2015.

Truax, Barry. *The World Soundscape Project's Handbook for Acoustic Ecology*. Vancouver, BC: A.R.C. Publications, 1978.

Underwood, Ted. The Stone and the Shell. Blog. Accessed January 13, 2018. https://tedunderwood.com.

Underwood, Ted. *Why Literary Periods Mattered: Historical Contrast and the Prestige of English Studies*. Palo Alto, CA: Stanford University Press, 2013.

Wagner, Laura. "Bringing Radio Haiti Home, One Step at a Time." H-net, June 29, 2016. https://networks.h-net.org/node/116721/blog/h-haiti-blog/132330/bringing-radio-haiti-home-one-step-time.

Wall, John N., PI. Virtual Paul's Cross Project. Accessed June 10, 2014. http://vpcp.chass.ncsu.edu.

Weheliye, Alexander G. *Phonographies: Grooves in Sonic Afro-Modernity*. Durham, NC: Duke University Press, 2005.

I | THEORIES AND GENEALOGIES

01

ETHNODIGITAL SONICS AND THE HISTORICAL IMAGINATION

RICHARD CULLEN RATH

I raised my hand, uncertain but determined. The professor, Africanist John Thornton, had asked if anyone knew music. I hesitated. It was 1988 and I was an adult scholar newly returned to school to pursue my BA. I was uncertain whether I was qualified for anything at that point. I had been playing guitar for about fifteen years and had a basic understanding of music theory, but I was not formally trained. But I thought, "I'm paying my tuition, so . . ." I raised my hand, narrowly beating out another student who later told me she hesitated a moment longer than I had. The decision set off a chain of events that profoundly affected my trajectory through life and career.[1]

The document Thornton gave me was just three or four photocopied pages from an old book that he had found on a research trip. The book was Hans Sloane's *Voyage to the Islands* (1707), a natural history of the islands off the west coast of Africa and in the Caribbean, where he focused most of his attention on Jamaica.[2] The pages in question contained a paragraph or so of text describing the music and dance taking place at a gathering of enslaved Africans on a Jamaican plantation in 1688; an engraving of a pair of stringed instruments and some vines used to clean teeth (and perhaps used as strings); and two pages of music described and transcribed by Sloane and his musician friend Baptiste that fell under three headings—Angola, Papa, and Koromanti.

Over the course of the next two-going-on-three decades, my interest in ethnographic history and the emerging field of sound studies was transformed by the introduction of digital sound to personal computers. This article traces that trajectory and its evolution into what I am calling "ethnodigital sonics," a term that emerged from a conversation with David A. M. Goldberg at the University of Hawai'i Digital Arts and Humanities Initiative over what it was exactly that I did with sound in my research and my music as well as in our collaborative work in the initiative.

"Ethno" refers to the expanded interdisciplinary approaches that ethnohistorians and ethnomusicologists follow to understand histories and musics that are otherwise somewhat incomprehensible through traditional single-disciplinary approaches. In this case, I have drawn on linguistics, history, anthropology, and musicology to arrive at conclusions not available had I taken any single approach in isolation. In contexts other than the one used in this chapter, I have employed this "ethno" approach to western subjects, so the label describes the approach, not who or what is studied. Uncertainty is inherent to this project, given that in this sort of work, the sum of the source material still adds up—according to ethnohistorian Patricia Galloway—to fragmentary, multiply biased, partially understood glimpses.[3] By making the "ethno" prefix characterize the method rather than the object of the study, I hope to bypass the justifiable criticism that ethnohistory replicates colonial power relations by offering different types of history for colonial actors than are offered for their "others." I think the methodological innovations of the approach are too substantial to warrant simply jettisoning the term. It shares this ecumenical approach with cultural studies and its related fields, so perhaps the prefix will end up being irrecoverable. I don't want to lean too heavily on it when the thing can be named in other ways.

While much ink has been spilled and many bits flipped on the subject of method in ethnomusicology, my reading has always been specific and goal-oriented: understanding a fragment of music or a snatch of transcription in the context of a particular time and place. For the African music in Jamaica project, my key sources from ethnomusicology are the foundational work of J. H. Kwabena Nketia on the music of Africa and Ken Bilby's groundbreaking work on the music of Maroons in Jamaica.[4] Although I am aware of the many limitations inherent in Western musical transcription, the fact remains that in history the fragmented glimpse is often all we get. I cannot, as one ethnomusicologist suggested, "go back out into the field" for more. Galloway's warnings to interpret cautiously and suspiciously and the historian's stance of uncertainty are the talismans here, since the questions do not vanish just because the methods are inadequate.

As for the "digital" component of the term, digital audio was slowly emerging as an accessible technology in the 1990s. The Musical Instrument Digital Interface standard (MIDI, introduced in the previous decade) became available, for better or worse, on every personal computer with a sound card, and it opened up new music-making opportunities along with the cheesy game tunes. By the mid-1990s relatively inexpensive full duplex sound cards came to market that brought the recording of CD-quality digital audio within reach on personal computers. Macs became the tool of choice for musicians, but I could never quite afford one, and PCs—first running Windows and then much later Linux—slowly caught up while offering more choice, complexity, and ways to go wrong at a cheaper price. Somewhat reluctantly, I became caught up in the latter two systems. By the end of the decade, audio-file compression made the storage and exchange of music feasible for professional sound artists and musicians, with the unintended side effect of setting off a revolution of sorts on the consuming end when the algorithms broke free.[5] In the first decade of the twenty-first century, digital synthesis and recording moved seriously into the realm of the personal computer with the maturing of consumer-priced digital audio workstations (DAWs) and the introduction of the VST and AU plugin formats that they hosted. Professionals also had another, costlier format for the Pro Tools DAW called RTAS. In particular, software brought samplers—a high-cost piece of hardware in the 1980s and 1990s—within range of any budget. By using a sampler, a MIDI pattern editor, and a player called a sequencer, I have been able to create and play instruments that would have otherwise been impossible on one or the other front.

I use the term "sonics" to signify the full range of thinking about, listening to, feeling, and making sound, including but not constrained to the field of sound studies as it emerged in tandem with these digital innovations.[6] Historians working within sound studies should note that hearing comes into play in two ways. First, hearing has a history: the senses are culturally and temporally shaped, and soundscapes of previous times are recoverable. One could take the paragraphs above as a personal, somewhat technology-driven version of this, as dozens (if not hundreds) of books and articles of wider scope have appeared over the past fifteen or twenty years. Second, we can hear history: that is, we can use our ears to understand the past, which is the topic of the remainder of this essay.

Early on in this journey, before sound cards were available and somewhat affordable, I puzzled obsessively over the few pages Thornton had given me and tried to imagine the sounds. I concocted instruments to test ideas, one of which I have kept around to this day. Putting my fingers to homemade or adapted sound makers and playing the written music made it clear that certain instruments were used on particular parts of the score—a kind of embodied sonic knowledge that I could not get from the text or images. In the Angola piece, the two-stringed banjo-like instrument shown in the engraving was played in the bass register because of the fingering it demanded. The upper register of that piece was almost certainly played on the eight-stringed harp, as it had eight notes in total, and they sounded somehow better when they rang out harp-like than when muted by left-hand fingering on a neck. (The upper register probably indicated the vocal melody as well.)

My task in interpreting Sloane's pages was to exercise what I call the historical imagination. I was not trying to reconstruct "authentic" performances, then or now. I was learning through the combination of touch and hearing that is fundamental to much music making. The interesting dynamic of the emic (roughly, the insider listening out) and the etic (the outsider listening in) came into play, since obviously I was in the latter, outsider position (which is of course my emic).[7] I was trying to imagine my way into the sounds and the history, not only from reading and thinking but also through doing and making. My idea was to make sounds using the principles that I thought the musicians Sloane and Baptiste used rather than create an "accurate" reconstruction, the latter a task too fraught to even begin. Some of the principles were aesthetic, others were implied by the constraints of Atlantic slavery, and one was discovered in the failings of the transcriber but not the transcription. A sampling of the principles would include: the choice of one scale over another in music deriving from a particular region; improvisation in the making of instruments using the materials at hand rather than a free selection; microtonal tunings, syncopation and polymeter; and choices about particular timbres.[8] While I can make no claims to have achieved insider understanding, I found much to value and learn from trying.

The music I have made for this project over the years requires as much imagination on the listener's part as on mine. I do not know how the night sounds and fire crackling in the background of my most recent digital attempts are made. Ultimately, I ended up getting closest to the sound of a live setting from an utterly artificial set of processes—sampling and sequencing—but that waited until the hardware and software had come within my reach. The setting on the Jamaican plantation in 1688 was obviously different from the conditions of reception, whether reading or listening, and layers of meaning exist in the distinction between audience and performer that would have been foreign to the Africans playing the music, although such distinctions are intrinsic to current understandings.

The stakes in these historical imaginings are high. For example, I am unwilling to take on the voice of enslaved Africans myself, and I am equally queasy about making it a singalong with audiences of mostly white folk providing the handclaps and the "Alla, Alla" refrain of the Angola piece. In light of the long history of minstrelsy (a tradition that perhaps lingers in the form of white suburban consumption of hip-hop), such a performance would adumbrate the power relations of both historical and present-day race relations and elide cultural appropriation into feel-good, irresponsible pop history. I think we can learn as much from what is left undone, unsung, and unplayed as we can from what is not, and I will not give voice to singing that linguistic and musical evidence conveys as having been hauntingly spectral, the voices of a community carried far from home to a strange and brutal land.

I was game to try the music, though, encouraged in that direction by the way musicians constantly borrow across cultures without the same sort of constraints that arise with vocal performance. I also could learn from what was absent as well as from what was there—drumming is made mostly of patterned absences, after all. The drums were missing altogether in the Sloane music. He reported that the musicians' use of drums "in their Wars at home in Africa" made them "too much inciting . . . to Rebellion, and so they were prohibited by the Customs of the Island."[9] In West African music, particularly that from the region that Koromanti designates, drums served as an immanent display of state power. They meant serious business in the Americas as well. Slave revolts, including a successful one in Jamaica a decade or so before Sloane's visit, were often organized around a drumbeat, sometimes a particular one recognized by the rebels as both a signal and a sign under which they fought for freedom.[10] Their absence is thus as meaningful and significant as the presence of other instruments.

I used a cheap nylon-string guitar to stand in for the two-stringed banjo-like instrument. I played the two top strings of it in a dropped tuning and wove a thin strip of torn paper between the strings to create a dull, buzzing sound, a trick I learned as a teenager when I wanted a fuzzy electric guitar effect and only had an acoustic. I wanted the buzzing for three reasons. First, the image showed an instrument with no bridge, which would make it sound muted and buzzy. Second, the aesthetics of much West African music value this buzzing, a preference that not coincidentally can be found in the fuzzy, distorted guitars in modern music from early electric blues onward. Third, the instrument in the background in the images of African music and Jamaica, used for comparison, is either a South Asian *tanpura* (which also has a flat bridge that imparts a characteristic buzzy and harmonically rich sound called *jivari*) or, alternatively, it is a Native American instrument made by forced-labor immigrants sold from the Carolinas into slavery in Jamaica at a rate of two enslaved indigenous people for one enslaved African.[11] Either is possible, since Sloane collected instruments from India and was an ardent comparativist in his study of natural history with proclivities toward the "cabinet of curiosities."[12] He just labeled the comparative instrument in the engraving as "Indian," so it is impossible to determine with certainty what he meant. I opted to emulate the South Asian instrument because of the buzzing.

The eight-stringed harp, which played the eight-note upper register in the Angola piece, was another adaptation of what I had at hand. This time I took two acoustic guitars (one nylon string and one steel string, since that is what I had), wove in the paper strips, tuned eight of the open strings to the notes of the upper register, set the guitars next to each other, and picked out the melody. This captured the open, sustained sound of a harp as well as the characteristic buzzing. Again, as a reconstruction I have no idea of how it fared—I like to think reasonably well—but as a tool for figuring out which instrument played which part, it was a useful exercise in historical imagination that helped me understand the music and musicians.

The Papa piece was too short to make much of, and the three pieces subsumed under the Koromanti title did not fall neatly into instrumental patterns when played on my emulations of the two-stringed banjo and the eight-stringed harp. For the Koromanti pieces I returned to Sloane's textual description of a musician playing on "the mouth of an empty Gourd or Jar."[13] Since the other instruments did not seem to fit these musical passages as well as in the Angola piece, I surmised, with no great certainty, that Sloane only saw the mouth of the gourd and the musician's hands and had missed that it held a *sansa* (thumb piano) that used the empty bowl as a resonator. This is speculation, and it is possible to play the Koromanti pieces on a modern banjo, but they become much more difficult on the two-stringed banjo because of the long ascending and descending passages, which are impossible on the eight-stringed harp.[14] The sansa's keys, which sound consecutive notes in a scale on alternating sides, facilitate exactly such ascending and descending runs.

I thought about the constraints the musicians were under. They had the knowledge and principles of the sansa but an impoverished access to materials and tools. What could they make with what was at hand? Probably nothing with the metal keys of modern African sansas—the ones usually known as mbiras—as metal would be valuable and scarce. Plus, the instrument would need to make a sound too distinctive to be mistaken for percussion. I guessed that they used thin strips of wood or bamboo as the keys to a sansa-like instrument.

I wondered how to emulate the sound. At the time, I was working at a shoe store to pay for tuition, rent, and food. New shoes often came stuffed with paper in the toe and a thin, eight- or nine-inch-long strip of flexible but sturdy bamboo that held the paper in place. I began saving them. I checked their sound by holding one end to the edge of a desk and plucking the other end. Lengthening or shortening the overhanging distance changed the pitch recognizably, but it still had a satisfyingly woody, buzzing, percussive thunk. I collected eighteen of the best-sounding ones and with a strip of wood trim pinned them to the bottom of an old dresser drawer that a previous tenant had left in the basement of my apartment. The drawer had a nice resonant sound. A slim-diameter piece of dowel jammed under the bamboo strips on one side of the wood trim made a bridge. I suppose if I had been attempting a reconstruction rather than practicing historical imagination, I would have waited for fall and gotten a big round gourd rather than a dresser drawer.

Sliding the bamboo strips forward and backward, I could adjust the notes of my sansa, which I dubbed the "bamboomba," to any scale I desired, including microtonal ones. This turned out to be quite important, because one of the three Koromanti pieces had an extra note in it that threw everything off kilter (fig 1.1). I guessed from playing around with the sansa that two of the notes in the Koromanti transcription actually indicated a single note that was a microtonal interval to its neighbors. Baptiste, the transcriber, could not parse the note through the filter of European notation, so he wrote the "off" note as two notes, flickering between them in his transcription. Changing the note to a microtone yields what is now called a "blue note," one that would fall between the notes of piano keys and a sound familiar to anyone who has heard blues or rock guitar.

I recorded my guitar and bamboomba experiments on a bottom-of-the-

FIGURE 1.1 The bamboomba.
PHOTO BY THE AUTHOR.

line four-track cassette recorder but was never particularly happy with the outcome: although the treated guitars taught me a lot, they sounded more like treated guitars than the instruments I imagined. In contrast, the bamboomba sounded just like I imagined it. Unfortunately, I could not put in the time to learn the three Koromanti pieces in any but the most halting manner.

I finished an article on African music in Jamaica in 1993 and set the project aside during my first few years of graduate school. By 1995 I had saved up for a real treat—an early "full duplex" sound card that allowed the simultaneous recording and playback required for multitrack recording. This launched the digital part of my journey in earnest. I could record up to four tracks, though only one at a time. My first project was to create a multitrack recording of the two-register Angola piece, using a guitar with two detuned strings for the lower register and the same guitar played on all strings for the upper. A third track was devoted to a simple percussion line.

The soundcard also had a nice on-board MIDI synthesizer so that I could write MIDI sequences and play them back. I could load my own sounds into the card too, but it took forever and was not worth the trouble. It did introduce me to sampling, though, which later would play a bigger role in my research when the technology became software based and fell within my reach in both price and ease of use. The General MIDI instrument specification provides for a sansa as one of its instruments, so even though it was a metal-keyed version that could not produce microtones, I programmed the Koromanti pieces in their full 1995 cheesy MIDI glory.[15] I shared my recordings with a few people who used them for teaching or were just interested, and they achieved a little *samizdat*-style circulation. They became part of a CD hypertext project called *Migration in Modern World History*.[16] When the MP3 format began to take hold, I posted them, along with many other pieces I had recorded, on my website and began using them in my teaching as soon as sound was incorporated into classroom computer setups.[17]

Sonic presentation clarifies sonic ideas. Learners do not easily grasp concepts like microtones and polyrhythm through language. They readily hear them though. Polyrhythm, for example, is much easier to teach through a simple clapping exercise in which one side of the room claps every third beat and the other side claps every fourth. Generally, some comedy ensues the first few tries, but then the audience hears it, and once that happens, they know it. And everyone has heard microtones in the bent strings of a lead guitar, at which point they cease to be something esoteric and become "oh that, of course." Having classroom computers with sound cards has had a transformative effect (though perhaps not for the clapping) on being able to teach with sound.

This type of learning pays off by making more accessible the experiences of people who are not well represented in traditional documentary sources. Sound delivers affect and the ability to strike the nonrepresentational aspects of being. When working to document the lives of enslaved Africans, I seek to avoid one of the oldest and most patronizing ways of telling the histories of people underrepresented or misrepresented in the documentary evidence: framing them as emotional and implicitly irrational beings, perhaps lacking in powers of representation, and presenting them as a sort of foil to logical, rational, and often paternalistic textual representations. Perhaps the most thoroughly articulated versions of this type of patronization are the theories of orality and oral culture that posit a great divide between literate thinking and that of everyone else. Recently, though, affect studies—which attend to nonrepresentational forces, the embodied relational abilities to affect and be affected—have come to the fore in new ways, so maybe there is yet a role for affective histories that are not reducible to hoary generalizations.[18] The practical part of sound's presence is that students and audiences connect to it in a different, perhaps more direct, way than to texts, a connection grounded in the body and experience as well as (and not in contrast to) the representational and the textual.

An oft-used trope in literary studies and historical work argues that since the source documents are biased to the core, we can tell nothing of the subjects they purport to describe, having only the psyche and fantasies of the white, usually male authors of the documents in question. Gayatri Spivak famously answered the question, "Can the subaltern speak?" (in reference to the practice of widow burning in colonial India) with a provisional, "Not really." She argues that despite the damages of colonial oppression, or in fact because of them, all we can really know about the widows is what the British authors of the texts thought they were experiencing—but nothing of the women's experience itself.[19]

In the Sloane materials, however, we have an interesting incursion. Sound is promiscuous, infiltrating and mixing freely without any attention to the intentions or desires of the listeners or even the producers. Sloane and Baptiste recorded the microtones, the polyrhythms, and the buzzing aesthetic without having a framework for doing so—the concepts were simply not present in Western music of their time. It was as if the Brahmin widows had been able to write through the mediumship of the British authors, who transcribed in a language they did not know and were not aware of writing. This hidden transcript in the Sloane materials only emerges when the sounds are rendered audible. While sounding them out does not offer unmediated access to the subaltern past (any more than do the texts for power holders), it does provide glimpses into the processes of cultural formation under the duress of slavery that move beyond models of resistance and accommodation, which are by definition always described in the terms of the master class.[20]

The problems of bias extend beyond the text into the digital domain as well. Ethnodigital musicians battle an implicit Eurocentrism in music software and hardware. It can be overcome, usually via a workaround or a deviation from an implied norm derived from Western classical music. For example, rendering polyrhythms using sequencers (composing software that is something like piano rolls) is easy in some applications but harder in others, depending on whether the bars of the sequencer can be set to different lengths. Many drum sequencers favor powers of two, having fixed lengths of eight or sixteen beats, neither of which is divisible by three, of course. For the simplest polyrhythmic work of three against four, the length of the sequence needs to be twelve or a multiple thereof, but this is often not an option. More complicated polyrhythms, which abound in non-Western music, are even less accessible. The bias, though not the rule, is toward a steady 4/4 meter, and the ethnodigital sonician has to improvise and make do in many cases.[21]

Creating microtones in MIDI presents different challenges. Part of the difficulty stems from the fact that MIDI is a Eurocentric language, the "nouns" of which are equal-temperament notes that must be bent and shaped by the "adjectives," in this case pitch-bending messages that modify the notes on an individual basis. Equal temperament is the norm, and microtones are temporary deviations. This can be countered in some software through tuning files that change the values of the "nouns" at the outset, mostly by adjusting the pitch-bending deviations behind the scenes. Until recently, though, the software has been fiddly and difficult or prohibitively expensive, and writing the tuning files remains a challenging enterprise. My goal of rendering MIDI files that sound more accurate, both in timbre and in tuning, took a long time and many false starts to realize.

Ethnomusicologists have dealt with these constraints in Western music notation for years, but they take on a different valence when they are built in to the composition process rather than the analysis of the music. It would be interesting to see what sort of musical language MIDI would have been if it had been designed by a consortium of musicians from around the world rather than a panel of music industry representatives. A newer protocol has emerged in recent years; called open sound control (OSC), it addresses some of the issues raised here from the ground up, but its adoption both in software and by musicians has been slow.

Despite this slow and uneven development, word about the musical versions I made of the Sloane transcriptions did circulate. Someone from a public television documentary team approached me in 2006 about using the recordings for a special on Jamestown and the first Africans who were sold there in 1619. I shared my (nonmicrotonal) MIDI files of Koromanti and the treated acoustic guitar rendition of the Angola piece. After some back and forth, the team decided not to use the music because "it was too upbeat for the portion that we were applying it to! Beautiful music—but we were looking for something a little more 'down.'" I was intrigued (and of course a little disappointed) by this rejection. I can see why a documentary that has a limited time to express a complex idea to people unfamiliar with it would not want any gray areas in the interpretation, especially when they could be construed as making slavery seem happy or as portraying the people within it as carefree, with no other thought in the world beyond their day off and some music and dancing. Such positions arose from the work of Ulrich Bonnell Phillips and other southern historians from the early twentieth century, and the "carefree slave" trope haunted popular culture well beyond the 1960s.[22] The documentarian's rejection of the music was a prophylactic reaction to this older, decidedly racist strand of American historiography, the critique of which has, over the course of eight or so decades, seeped into the public consciousness at least as far as the television documentary.

But the sound and affect of the Angola and Koromanti music also raise an interesting question: Is there such a thing as absolute slavery? Can the soul of another human being be controlled so absolutely that we can reduce that person's emotional range to nothing but sad songs and sorrow? I like to think of music as a release, an autonomous zone apart from the usual tropes of resistance and accommodation. I could sound the Sloane excerpts not necessarily in reference to slavery, but not necessarily discounting it either, attributing more to the human than just "the slave." The musical pieces I was studying made a space where enslaved Africans could be sometimes fierce, sometimes joyous, sometimes lustful, sometimes sorrowful, and more fully human than was possible within the constraints slave masters both imagined and tried to realize. Here were several competing historical imaginaries: those of the Africans, the masters, Sloane, the television documentary, Ulrich Bonnell Phillips, the audience, mine, and yours, as well as those of other audiences with whom I have shared this idea.

I think that the idea of total slavery with absolute slaves was ultimately the fantasy of the masters rather than a description of lived experience. The enslaved never became the comprehending-but-not-thinking working objects that their masters wished them to be, as brutal as the institution was. Nor were they all and only "down." I would posit that the full range of emotions that can be culled from these pieces—everything from the ghostly sorrow of lost and wandering human communities to "too upbeat"—is evidence of such. The affective power of sound to change the ways people experience things even before and long after they have thought about them means that it matters.

This question of the destructive effects of slavery on African culture brings us back to an old but fascinating debate initiated in 1939 between the African American sociologist E. Franklin Frazier, who made a case for the utter destructiveness of slavery on a usable African past, and the white anthropologist Melville Herskovits, who argued the contrary position by cataloging hundreds if not thousands of Africanisms in American culture. Frazier got the better of the argument well into the 1960s. It became important enough that in 1965 Daniel Patrick Moynihan, an adviser to President Johnson at the time, framed it as his infamous "culture of poverty" argument, which became the basis of welfare policy for the next three decades. The Black Power movement shifted the momentum to Herskovits's side, and responding to this politicization, anthropologists Sidney Mintz and Richard Price proposed a mediating approach using creolization as their theoretical frame. Unfortunately, this has more often than not meant a return to the Frazierian claim touting the destructive and innovative effects of slavery at the expense of African culture, so much so that some historians have mistakenly situated creolization in opposition to cultural continuities in the Americas, treating them as competing and antagonistic models.[23]

What the music points toward is a more nuanced understanding of how all these tendencies—creation, destruction, and persistence—were and are integral to each other for anyone who thinks deeply about today's multi-billion-dollar popular music industry. All are necessary ingredients to understanding the complexities of African—and European—life in the Americas. This understanding of creolization shows us how cultural continuities, destruction, and innovation were twisted together in a highly creative generative process that continues to have effects to this day.

But maybe it was the microtones. I found that time and new, less expensive sequencing and sampling software made it possible to play the missing microtones, even when I was not sufficiently adept to play the music on the sansa I had made. I tuned the bamboomba to Western scales and recorded for each of the keys a soft, medium, and loud pluck. I imported these into a sampler that could trigger the appropriate sample when notes were played, and I massaged the notes in the sampler to play the microtonal scales I needed (fig. 1.2). I am no better on keyboards than I am on sansa, so I tried triggering the MIDI notes through a guitar with a MIDI output on it. This was moderately successful, and I went on a mini tour of two conference performances in 2010 and 2012. The setup was awkward, bulky, and skittishly complex to prepare for a performance. After a particularly difficult second outing, I have put the live version on hiatus (unless you want to book me!), but it is quite fun to play a guitar and have a completely different instrument emerge from the speakers.

A better solution was to sequence the MIDI—I had already done this—and then use the tweaked samples to capture the microtones. Where Baptiste's flickering notes occurred, I resolved them into a single microtonal note. This meant I could digitally separate what was played from the notation, which better resembles what I imagine the process to have been. After years

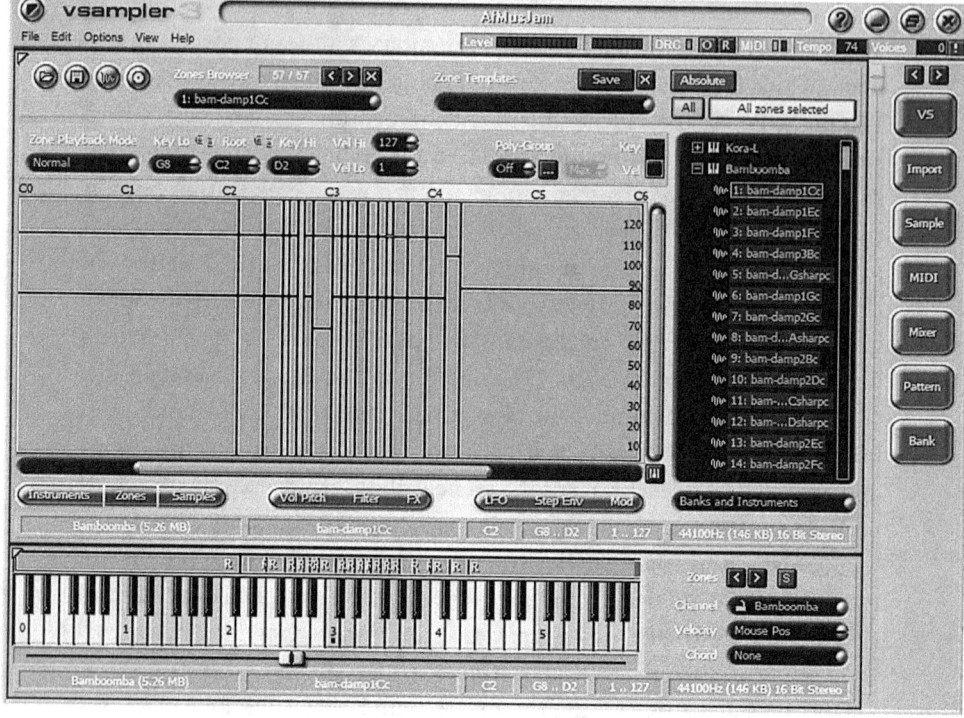

FIGURE 1.2 Recording of the bamboomba imported into a sampler.

of pondering the problem, I have a solution and a version—minus vocals and plus some night sounds and percussion—that approximates what I imagine the instruments and dancers sounded like on that night in Jamaica in 1688.

Using digital tools and a bit of historical imagination, both mine and my listeners', I have found inroads into understanding a bit more about the lives of Africans in seventeenth-century Jamaica as well as the processes, constraints, feelings, and creativity that went into building a distinctively new culture in the Americas under unimaginably harsh conditions. Although the trail of documentary evidence on this subject gives out if we limit ourselves to reading the texts, by setting aside the academic perquisite of *ex cathedra* certainty, whole new avenues of understanding open up when we listen. I have tracked my path through this process as constituting "ethnodigital sonics," and I would offer that path as one way of undertaking the practice of it. I hope others will find this a useful road map for the practice and its possibilities.

NOTES

1. The most recent versions of my performances of the musical pieces referenced in this chapter are available at "Ethnodigital Sonics Meets Maker Culture in Seventeenth-Century Jamaica" (accessed January 16, 2018, http://way.net/AfMusicJam).
2. Sloane, *Voyage to the Islands*, vol. 1. As I was revising this essay, an excellent new website that highlights the section of Sloane under consideration in this chapter came online: see Dubois, Garner, and Lingold, Musical Passage: A Voyage to 1688 Jamaica.
3. Much along the same lines found in Galloway, *Practicing Ethnohistory*, 27, and throughout.
4. Nketia, *Music of Africa* and *African Music in Ghana*; and Bilby, "Kromanti Dance," "Caribbean as a Musical Region," and "How the 'Older Heads' Talk."
5. Sterne, MP3.
6. The best single-volume introduction is Sterne, *Sound Studies Reader*.
7. For the dynamic approach to the terms I have taken, see Hymes, "Emics, Etics, and Openness," along with the other essays in that volume.
8. For the details, see Rath, "African Music in Seventeenth-Century Jamaica," and *How Early America Sounded*, 8–9, 68–93.
9. Sloane, *Voyage to the Islands*, 1:xlviii–xlix, lii.
10. Rath, "Drums and Power."
11. Gallay, *Indian Slave Trade*.
12. On cabinets of curiosities, see Kupperman, *Indians and English*, 21–22, 349n.13; for the specific context of Sloane's collection (which became the basis of the British Museum), see Delbourgo, "'Exceeding the Age in Every Thing.'"
13. Sloane, *Voyage to the Islands*, 1:xlix.
14. For a video of the Koromanti piece played on banjo, see Burton, "Older than Minstrel." David K. Garner plays several of the Sloane pieces on a fretless banjo on the Musical Passage website cited in note 2.
15. MIDI itself has no sounds, only instructions to tell a synthesizer to sound a certain note at a certain pitch and velocity (relative volume) for a certain time on a certain channel from a certain sound bank. The sound banks of the synthesizer can contain any arbitrary MIDI-capable synthesizer, including the ones built in to Microsoft and Apple operating systems. "General" MIDI is a specification implemented to bring some predictability to the sounds a MIDI file produces and includes a numbered set of 128 target instruments, of which number 109 in the "Ethnic" group is the kalimba, a metal-keyed sansa. The actual sound produced is still left up to the synthesizer, which used to be

a separate board on early soundcards. In the 2010s, as computers have grown more powerful, the synthesizer on a home-use personal computer is generally executed in software. The sound itself thus was built into my sound card in 1996, but it was specified by General MIDI program 106, hence the awkward wording of the passage. For details on General MIDI, see MIDI Manufacturers Association, "About General MIDI."

16   Manning et al., *Migration in Modern World History*.
17   Rath, "African Music in Seventeenth-Century Jamaica."
18   Gregg and Seigworth, *Affect Theory Reader*, and Goodman, *Sonic Warfare*.
19   Spivak, "Can the Subaltern Speak?" For an important rejoinder, see Mani, *Contentious Traditions*.
20   For hidden transcripts, see Scott, *Domination and the Arts of Resistance*. For alternatives to the resistance models and the ways the cultural formations took shape in North America and the Caribbean, see Rath, "African Music in Seventeenth-Century Jamaica," "Echo and Narcissus," and "Drums and Power." For hegemony, see Gramsci, *Selections from the Prison Notebooks*, ably critiqued in Lears, "Concept of Cultural Hegemony." A whole body of slavery studies from the 1950s to the present is based on the notions of resistance and accommodation but is beyond the scope of this chapter to review.
21   An internet search for "Euclidean Rhythms," itself a somewhat Eurocentric framing, will return a host of links to wonderful homemade software based on the research of Godfried Toussaint, but the inventions take the form of optional extensions to the major music creation platforms rather than being integral to any of them. See Toussaint, "Euclidean Algorithm." One iOS app, Patterning, from Olympia Noise Company, does manage to incorporate polyrhythmic possibilities by using a circle rather than a piano-roll model, with each of the rings that constitute the sequencer divisible into any number of steps.
22   Phillips, *Life and Labor*.
23   The Herskovits-Frazier debate is attended to in depth in Rath, "Drums and Power." For the mistaken notion that creolization is opposite to African continuities, see Lovejoy, "African Diaspora."

WORKS CITED

Bilby, Kenneth M. "The Caribbean as a Musical Region." In *Caribbean Contours*, edited by Sidney W. Mintz and Sally Price. Baltimore: Johns Hopkins University Press, 1985.

Bilby, Kenneth M. "How the 'Older Heads' Talk: A Jamaican Maroon Spirit Possession Language and Its Relationship to the Creoles of Suriname and

Sierra Leone." *New West Indian Guide* 57, nos. 1–2 (1983): 37–88. https://doi.org/10.1163/13822373-90002097.

Bilby, Kenneth M. "The Kromanti Dance of the Windward Maroons of Jamaica." *New West Indian Guide* 55, nos. 1–2 (1981): 52–101. https://doi.org/10.1163/22134360-90002118.

Burton, Janet, et al. "Older than Minstrel, Video Added with All 3 Parts—Discussion Forums." *Banjo Hangout*, July 4, 2012. www.banjohangout.org/archive/239951.

Delbourgo, James. "'Exceeding the Age in Every Thing': Placing Sloane's Objects." *Spontaneous Generations* 3, no. 1 (January 2010): 41–54. https://doi.org/10.4245/sponge.v3i1.6743.

Dubois, Laurent, David K. Garner, and Mary Caton Lingold. Musical Passage: A Voyage to 1688 Jamaica. Accessed November 5, 2017. www.musicalpassage.org.

Gallay, Alan. *The Indian Slave Trade: The Rise of the English Empire in the American South, 1670–1717*. New Haven, CT: Yale University Press, 2002.

Galloway, Patricia Kay. *Practicing Ethnohistory: Mining Archives, Hearing Testimony, Constructing Narrative*. Lincoln: University of Nebraska Press, 2006.

Goodman, Steve. *Sonic Warfare: Sound, Affect, and the Ecology of Fear*. Cambridge, MA: MIT Press, 2012.

Gramsci, Antonio. *Selections from the Prison Notebooks of Antonio Gramsci*. London: Lawrence & Wishart, 1971.

Gregg, Melissa, and Gregory J. Seigworth, eds. *The Affect Theory Reader*. Durham, NC: Duke University Press, 2010.

Hymes, Dell. "Emics, Etics, and Openness: An Ecumenical Approach." In *Emics and Etics: The Insider-Outsider Debate*, edited by Thomas M Headland, Kenneth Lee Pike, and Marvin Harris, 120–26. Newbury Park, CA: Sage Publications, 1990.

Kupperman, Karen Ordahl. *Indians and English: Facing off in Early America*. Ithaca, NY: Cornell University Press, 2000.

Lears, T. J. Jackson. "The Concept of Cultural Hegemony: Problems and Possibilities." *American Historical Review* 90 (1985): 567–93.

Lovejoy, Paul E. "The African Diaspora: Revisionist Interpretations of Ethnicity, Culture, and Religion under Slavery." *Studies in the World History of Slavery, Abolition and Emancipation* 2, no. 1 (1997); n.p. Accessed November 5, 2017. www.yorku.ca/nhp/publications/Lovejoy_Studies%20in%20the%20World%20History%20of%20Slavery.pdf.

Mani, Lata. *Contentious Traditions: The Debate on Sati in Colonial India*. Berkeley: University of California Press, 1998.

Manning, Patrick, Northeastern University, and World History Center. *Migration in Modern World History, 1500–2000*. Belmont, CA: Wadsworth, 2000.

MIDI Manufacturers Association. "About General MIDI." Accessed May 20, 2015. www.midi.org/techspecs/gm.php.

Nketia, J. H. Kwabena. *African Music in Ghana*. Evanston, IL: Northwestern University Press, 1963.

Nketia, J. H. Kwabena. *The Music of Africa*. New York: Norton, 1974.

Phillips, Ulrich Bonnell. *Life and Labor in the Old South*. Boston: Little, Brown & Co., 1929.

Rath, Rich. "African Music in Seventeenth-Century Jamaica." *Way Music*, January 16, 2008. http://way.net/waymusic/?p=13.

Rath, Richard Cullen. "African Music in Seventeenth-Century Jamaica: Cultural Transit and Transition." *William and Mary Quarterly* 50, no. 3 (1993): 700–26.

Rath, Richard Cullen. "Drums and Power: Ways of Creolizing Music in Coastal South Carolina and Georgia, 1730–1790." In *Creolization in the Americas: Cultural Adaptations to the New World*, edited by Steven Reinhardt and David Buisseret, 99–130. College Station: Texas A&M University Press, 2000.

Rath, Richard Cullen. "Echo and Narcissus: The Afrocentric Pragmatism of W. E. B. Du Bois." *Journal of American History* 84 (1997): 461–95.

Rath, Richard Cullen. *How Early America Sounded*. Ithaca, NY: Cornell University Press, 2003.

Scott, James C. *Domination and the Arts of Resistance: Hidden Transcripts*. New Haven, CT: Yale University Press, 1990.

Sloane, Hans. *A Voyage to the Islands of Madera, Barbados, Nieves, St. Christopher and Jamaica*. 2 vols. London, 1707.

Spivak, Gayatri Chakravorty. "Can the Subaltern Speak?" In *Marxism and the Interpretation of Culture*, edited by Cary Nelson and Lawrence Grossberg, 271–313. Urbana: University of Illinois Press, 1988.

Sterne, Jonathan. *MP3: The Meaning of a Format*. Durham, NC: Duke University Press, 2012.

Sterne, Jonathan, ed. *The Sound Studies Reader*. London: Routledge, 2012.

Toussaint, Godfried. "The Euclidean Algorithm Generates Traditional Musical Rhythms." In *Renaissance Banff: Mathematics, Music, Art, Culture*, edited by Reza Sarhangi and Robert V. Moody, 47–56. Southwestern College, Winfield, KS: Bridges Conference, 2005.

02

PERFORMING ZORA

Critical Ethnography, Digital Sound,
and Not Forgetting

MYRON M. BEASLEY

It is evident that the sound-arts were the first inventions and that music and literature grew from the same root.
— ZORA NEALE HURSTON, "Folklore and Music"

Research is formalized curiosity. It is poking and prying with a purpose.
— ZORA NEALE HURSTON, *Dust Tracks on a Road*

I just wanted people to know what real Negro music sounded like.
... Was the real voice of my people never to be heard?
— ZORA NEALE HURSTON, "Folklore"

TRACK 1: (Re)mixing Zora at the Rooster

The Red Rooster is more than just a restaurant on Lenox Avenue in the heart of Harlem. It could be mistaken for a library, an archive, a museum, or even a plush humanities center on a college campus. The primary wall is filled with books, magazines, albums, and other ephemera of black cultural production with the opposing walls well curated with visual art by noted African American artists. The bar sits in the center. On one particular Thursday evening the DJ huddles in the corner mixing the tunes with a turntable connected to his Mac Powerbook. "Birds flyin' high, you know how I feel," the first line from Nina Simone's hit "Feeling Good" (1965), permeates the restaurant.¹ Simone's strong, robust voice is reframed with a techno beat yet sustained with the slower moves of the most popular version. As the techno line crescendos into a clash, waves of vocal tracks disrupt the seemingly haphazard sounds. First the voice of Langston Hughes reading "I have the weary blues," followed by the stern, firm voice of Maya Angelou vocalizing lines from I Know Why the Caged Bird Sings. Followed by (probably the least familiar to casual listeners) a track that jarred most—the voice of Zora Neale Hurston crooning "Halimuhfack."² Hurston's track halted the flow. The gritty, dusty, scratchy quality transported the listener to another time, to another place. The pastiche of sounds curated by the DJ was a "remix" for sure. The audio palimpsest—layering a mix-match of music weaving literature, ethnography, biography, and history—the techno-digital moves of the DJ announced the confluence of digital sound and humanistic inquiry.

The performance at the Red Rooster reflects what cultural critics Martha Buskirk, Amelia Jones, and Caroline Jones proclaim in their review of the cultural productions of 2013: that this moment of humanistic inquiry is dominated with the prefix "re-."³ In this moment in the U.S. academy, the demise of the humanities is announced often in the popular presses, with some faculty who occupy such locations bewildered and clamoring for survival. Funding is shrinking and some departments are being eliminated. The justification for such measures is wrapped in the discourse of precarity. While many scholars feel the need to "resuscitate" the humanities, others take solace in "rethinking" and advancing and enhancing humanistic inquiry by reimagining with technology. The DJ spinning at the Red Rooster was a performance of reimagining humanistic inquiry and thus creating new forms of text—a haunted text evoking an infectious performance through the weaving of sound, technology, and digital elements.⁴ While Buskirk and her colleagues "re-create, reanimate, recast, recollect, reconstitute, reconstruct, reenact" in their review, rememory or recalling does not appear.⁵ As this chapter considers the sonic work of Zora Neale Hurston in light of the contemporary conversations regarding sound studies and digital humanities, it recalls the significance of Hurston in the changing domain of ethnography, thus situating performance as both a method of inquiry and an embodied phenomenon, the move with which performance spawns new forms of texts and modes of performance. Yet discussing black cultural production within the panoply of technology warrants a contemplation of capitalism, cultural meshing, and cooptation.

Alexander Weheliye reminds us, as did Henry Louis Gates (and others) before him, of the rambunctious nature of African American cultural production.⁶ The sense making of daily life within communities of the African diaspora manifests itself in a motley assemblage of performative acts. The means of documenting, representing, and preserving life were rarely confined to the printed text. Rather, the visceral experiences were *performed*—through orality. (I place material culture as performance.) What Gates locates as the "trope of the talking book," he adroitly explicates as the dialectical strains between the written and oral text. The violent history of literacy for blacks in the Americas is an exclusionary practice and presents a unique dynamic of engaging with the domain of black humanistic inquiry. Gates recapitulates the polyvocality and the orality that surround "texts"—paraperformance that lurks outside and around the written and oral "texts" that signal "always more than what it appears to be," or more familiar "signifying."⁷ Such playing with texts speaks to the fluid, infectious, and contested nature of black cultural production. Yet Weheliye, moving from Gates and others, contemplates the role of digital technology in "recorded" production and situates the deployment of black culture with the interplay between consumption (capitalism) and subject/identity discursive formations. At stake is erasure. Playing with the concept of "sonic Afro-modernity," Weheliye considers the rise of the phonograph and the reproduction and distribution of African American sound. As black sound attains "market value," the propensity for fetishization—a subject without citation, a subject without identity—becomes more acute. The commodification (read: capital) of black diasporic culture without critical interventions encourages the separation of the I am I be—subjectivity and identity—which Weheliye claims is the pervasive philosophy that has "run amok" in the Anglo-American humanities.⁸ To insist on the I am (subjectivity) I be (identity) as a unit, a both/and, a synergetic dialectic, allows for and opens up a space to foreground the sonic discourse in black cultural studies and provides more diverse ways to think more broadly about black cultural production. The audio palimpsest performed by the DJ at the Red Rooster is a fitting example of ways in which sonic imprints insert the past in present, lived realities—always recalling, never allowing a forgetting, but producing new forms of representations and suggesting new ways of engagement.

The tracks of Hurston's voice in the stream of a techno dance mix at the bar in Harlem were a "hailing" to be sure, but they were also a haunting. The domain of performance rests in its ephemeral nature. Once a performance happens it disappears, according to Peggy Phalen, suggesting ontology of disappearance.⁹ Derrida proposes (and I agree) a move beyond ontology toward a hauntology, as he advocates the nature of truth as derived from engaging the thing and not merely the thing itself.[10] Hauntology therefore highlights the persistent, contested, and infectious nature of performance.[11] Barbara Browning likens the generative nature of performance to infectious rhythms. The term vibes with Paul Miller, who like Browning uses epidemiology as metaphor to describe the ways in which digital technology has enhanced and, I dare say, reframed the work of a DJ (whom he labels a rhythm scientist).[12] For Miller, DJ mixes produce vectors that are capable of infecting agents that have the potential of becoming infectious.[13] Performance theorists are not preoccupied with the performative act itself but rather the generative nature that comes from engaging with the performance (performances never end), its mixings and ability to spin further questions, deliberations, and theoretical discourses. Like the DJ spinning and weaving different strands of cultural elements, digitized and meshed with technology, performance in its infectious nature flows and morphs into new forms of performances, new ways of participation.

As the consummate performance ethnographer, Zora Neale Hurston embraced the convergence of performance as a method of inquiry, exhibited the domain of performance as an embodied phenomenon, and exacerbated the critical space in between. A haunted space, lacuna of possibilities, also allows for the explorations of Hurston's work in the frame of the synchronous *I am I be* and other multiple interactions and perspectives, and creative, imaginative, and critical inquires.

The performance at the Red Rooster—the DJ moving, weaving Hurston's voice with manipulated recycled contemporary tones as the audience engaged in the happenings of the moment—was a creative and imaginative opening produced by the collaboration between sound and technology. The digitized voices of literary figures and the use of technology to mix and infuse diverse sounds make it possible for me to consider Hurston's work. Zora Neale Hurston *was* at this moment at the Red Rooster, on Lenox Avenue in Harlem, a site Hurston inhabited years before. The digitized sounds of her voice haunt us, creating more performative spaces of possibility, the chance to reimagine her and her work in different ways to different audiences, yet they recall her contributions and resignify her influence as to not forget her!

TRACK 2: Meeting Baldwin, Meeting Zora

On December 1, 1987, the voice on the radio in my small flat in Paris announced the death of James Baldwin: "James Baldwin, the black American writer... the author of *Go Tell It on the Mountain*." The announcement would continue throughout the day, as Baldwin was most revered in France. I had not known of James Baldwin, nor had I read any of his work. I rushed to the used bookstore down the street to search for work by this famous black American whom I did not know. I found a copy of *Go Tell It on the Mountain* on a bookshelf, but on the floor, just underneath the lower overflowing shelf, was a tattered copy of Hurston's *Their Eyes Were Watching God*. I purchased both.

I struggled at first while I silently read through the brown and occasionally soiled pages that disguised the breadth of knowledge that would come to have a significant influence on my life. I began to read her words aloud—carefully moving my tongue and lips to imitate the unfamiliar diction of the written text. The tongue slithered about, arching to touch the awkwardly placed plosive, but eased into the slowly paced diphthong. An enlivened spoken word captured me the reader, coerced me to listen to the nuanced sonic movement of her text.

At the death of Baldwin, I met Zora.

TRACK 3: Boas, Boas, Boas

When anthropologist Franz Boas asked his then-student Zora Neale Hurston to travel to Florida and record the folklore of her childhood town of Eatonville (1935, 1938), Hurston returned to Boas with audio of herself singing the folkloric songs of her and the city's past. This moment with Boas is significant. The performance by Hurston signaled an epistemological shift in the social sciences and humanistic inquiries surrounding how to "read," "write," and "represent" culture.

When I first encountered Zora Neale Hurston in 1987, I was a second-year student in college; my majors were oral interpretation/rhetoric and anthropology. The field of rhetoric in the discipline of communication studies then privileged the concept of "speech" or oratory and oral interpretation. Oral interpretation is the art of "suggestion" (as opposed to action) with the aim of, according to Charlotte Lee and Timothy Gura, communicating a text in its "intellectual, emotional, and aesthetic entirety" in words.[14] Students mastered the International Phonetic Alphabet, learned how to dissect the paralinguistic qualities of words, and scrutinized texts to enliven the written work for the ears. "Great literature was not only written to be read, it was also written to be heard," was a common refrain. Pedagogy and scholarship were preoccupied with the persuasive use of and performance of written texts.[15] Simply, oral interpretation was about the criticism of and the (re)performance (reading aloud) of written text (including poetry, speeches, and other forms of printed texts). As traditional anthropologists continued to stake claim to the word "culture" despite the emergence of feminist theory and ethnic studies in the academy, ethnographers advocated for greater inclusion of physical participation and self-reflection to develop theory and analysis. At the point of my matriculation to college, a theoretical shift was occurring. Modernism was being condemned and, as postmodernism was just getting its footing and stride, critical theory/cultural studies invaded and challenged contemporary thought in multiple ways, including oral interpretation/rhetoric and anthropology.

The primacy of the written text was debated, and the mere definition of "text" was challenged. Barthes proclaimed that the death of the author endorsed the power of the text itself, while Dwight Conquergood encouraged those in the speech/oral interpretation/rhetoric arenas to understand the cultural politics of the primacy of the written text as patriarchal and exclusionary.[16] Conquergood theorized the shift from reciting "literary" written texts to a performance of narrative—bodily stories—moving the domain of ethnographic inquiry to participatory engagement, critical intervention, and performance ethnography.[17] The move also challenged the presentation of scholarship, refocusing beyond the printed monograph to also consider the recitation of field notes, the performance of participants' interviews, and a reflexive turn to include narratives of "doing" the research. And even more significantly, the recognition of the fluidity of "power" and its manifestations in fieldwork experienced profound change. In the domain of researching the "other," scholars shifted emphasis from subjects to coparticipants, coresearchers, and, derived from Zora Neale Hurston, cowitnesses to the documenting, moving, and making of culture.[18] The participants became speaking subjects, not objects being spoken for. Oral interpretation moved from text to performance.

When Hurston performed the folklore from her fieldwork to Boas, she was already at this epistemological, ontological, and even methodological moment. Furthermore, the embracing of Hurston's performance embodied the archive, a topic recently interrogated by performance theorists as they contemplate other depositories of knowledges and histories. Verne Harris is preoccupied with questions about archiving bodily sounds: What does it mean to archive unwritten languages, ritual songs, and chants? He resists the concept of a physical, centralized holding place (i.e., archive) and instead endorses the continued visceral transmission of cultural variables through the teaching of such cultural performances through technology.[19]

In that moment with Boas, Hurston disrupted the binary opposition that plagues academic discourse by eliminating boundaries between the scholar and the participants and making known the cultural politics of doing fieldwork and producing creative and accessible ways of (re)presenting scholarship and creating new texts. The use of technology allows scholars access to different raw materials to develop the creative, physiological, and visionary texts of interrogation. The definition of the griot becomes broader. The DJ at the Red Rooster spins.

TRACK 4: Black Women Performing Blues

She could hold a tune in the shower peck out a few bars on
the piano and strum some decent chords on the guitar,
but she was no maestro.
— VALERIE BOYD

You'd most likely be hard pressed to find anyone who would
call her a great singer.
— DAPHNE BROOKS, describing Hurston's singing

Zora Neale Hurston is considered a member of the "unholy trinity," along with Billie Holiday and Bessie Smith. Angela Davis and Norman Denzin locate the genre of blues music as performed ethnographic text.[20] Performative texts are credible scholastic endeavors that articulate conditions of race and gender politics (in a form accessible to many); these performative arguments are based in lived experience and grounded in what black feminists label theories of the flesh.

According to Alice Walker, "Zora *belongs* in the tradition of black women singers. . . . She followed her own road, believed in her own gods, pursued her own dreams, and refused to separate herself from 'common' people."[21] The lyrical locutions of Holiday, Smith, and Hurston document and announce the unaltered reality of black life. The effusive performances that graphically call out love, sexuality, and even violence demonstrate the "intellectual independence and representational freedom" in the work of these great women.[22] Such performances were never simply about aesthetics, although they were great performances indeed, but there is always slippage—the messiness that bespeaks the extemporaneous nature of the blues. Never performing the same song the same way allowed for audience participation. The blues are always off-kilter, always mirroring the ebb and flow and nuances of everyday life. When I listen to Hurston's recordings—from the audio expedition with Alan Lomax and the work compiled during her stint with the Works Progress Administration (WPA) with Herbert Halpert—it is the imperfections, the background noise, the inserted questions, and her off-toneness that provide a fuller and more complete set of "data." They shape an integral part of the performances that reflects both the nuanced nature of ethnographic work and the nuanced nature of everyday life.

Hurston's work as a scholar of black life is not only significant ethnographically, in that she embraces "studying" her own community, but it also records a period of migration of blacks to the north and west at a time when immigrants and the growing class of educated African Americans were leaving "their downward, down-home ways and traditions behind."[23] She writes that collecting folklore "would not be a new experience for me. [W]hen I pitched headforemost into the work I landed in the crib of negroism."[24] Hurston recognizes the magnitude and importance of archiving and documenting such work. Yet she claims, "It was only when I was off in college, away from my native surroundings, that I could see myself like somebody else and stand off and look at my garment. Then I had to have the spy-glass of Anthropology to look through at that." She later disrupts the concept of "spy-glass" for her concept of feather-bed resistance. Recalling the "speakerly text," Hurston reveals how the black communities would engage in strategies of speaking-but-not-really-speaking to outsiders "coming to get information." A feather-bed resistance, she says, is when "we let the probe enter, but it never comes out. It gets smothered under a lot of laughter and pleasantries."[25] Maybe Hurston used her "insider" status to probe, or maybe she used her ability to participate in the community, to fully embody the experiences of many of her cowitnesses. Her method as performance insisted *I am I be* as a unit.

The Federal Writers' Project (FWP) was a program within the WPA, created by the Roosevelt administration to stimulate the U.S. economy during the Great Depression. One aim of the FWP was to preserve and document American folklore and traditions. The state of Florida folklore section was not established until 1939; however, Hurston had already participated in two folklore audio expeditions with Lomax in 1935 and 1937.[26] In 1939 she became an official member of the state of Florida's FWP committee that was officially titled the Joint Committee on Folk Art's Southern Recording Expedition. The audio materials were recorded on acetate disks spinning at 78 or 33⅓ rpm. The irregular speeds of the recordings are perhaps due to the heat incurred during travel from Florida to Washington, D.C. In addition, the scratches of the tapes featuring Hurston's voice influence the playback quality of the sonic performance. The Florida Memory Project currently houses and provides access to some of this work, but the Library of Congress holds most of the audio from the FWP.[27]

Hurston's recordings are nuggets of academic materials, filled with bits from African American life in the South, including information about labor and the economic, religious, and social lives of southern blacks. The sounds include Hurston singing and performing chants (mostly of African American railroad workers) and other speakerly, performative folkloric texts. For example, in the C recording, Hurston contextualizes the song and she performs it back to her cowitnesses. She is asked by audience members to share more about the lyrics. You hear the intermittent ideas and questions posed and then you hear her voice. Her brash utterance discloses her methodology. Someone asks, "Who taught you this song?" She replies, "Not one person." Rather, she discloses how she "would sing along with the crowd and then perform it back to them to make sure [she] had it correct." Representation is important to performance ethnography. Contemporary performance ethnographers return to the communities in which they work to ascertain if the final product (play, article, image, video) truly represents who the subjects are. Yet Hurston resisted her training with Boas, who was adamant about recording "the other," and the act of recording herself radically reframed the topic of representation. Hurston refused to translate the other through technology: instead we get her—Hurston the anthropologist, folklorist, interlocutor, and also Hurston the community member familiar with lives and culture of this particular population. The aim in Hurston's ethnographic reality is not authenticity but a realization of the contested nature of doing and the mere presentation of ethnographic inquiry. The digitized audio collection of the Florida expedition is an invitation to a dialectic aural performance, yet it highlights some of the challenges as to how to engage with Hurston's audio work.

In "Let the Deal Go Down" we get a sheer sense of Hurston's embodiment. In this song about gambling (connected to the card game "Georgia skins"),

Hurston rehashes the folkloric tale, and then she explicates how the song is literally performed as she embodies several characters sitting around a table to give the audience a real sense of the context. The listener surmises the interplay of community and economics in the rural South. Yet in "Let's Shack," the arch of Hurston's voice as she emits the hard "HAAA" and the short, abrupt phrases eclipsed by hard constant sounds evoke the hardship of labor on the railroads. I should also note that Hurston desired to document every aspect of black life. Some recordings from FWP and the expeditions with Lomax are not accessible online because of the bawdy subject matter. The digital is a contested space. Digital conversions do not exclusively guarantee accessibility. Even at this moment of collusion between reality and technology (simulacra), the cultural politics of technology, particularly in the context of the United States, continues to struggle with power. Who has the right to decide what is acceptable, what is scholarship, and what relevance certain materials hold? In addition, exclusionary politics ensures that everyone does not have access to digital materials. The performative nature of the sound—including the meta-analysis that Hurston herself provides, her visceral embodiment, and her attempt to make the work assessable in a variety of forms—positions her work in the synchronous *I am I be*, not separating the object from the subject.

Earlier in this chapter I revealed that my introduction to Hurston was through the paralinguistic oral qualities of her written texts, a writing that evokes the essence of sounds through embodied sensual performances. But it was the digitized sound recordings from the expeditions that profoundly influenced my academic work, from my ethnographic fieldwork (Brazil/ Haiti) and my research topics (ritual performance) to the presentation of my scholarship (installation, plays, sound). As a researcher, I find the area of digital sound studies considers even more ways to critically interrogate Hurston's digital work, particularly in the areas of scholastic presentation along with gender and race.

Hurston's audio work is scholarship. Her ability to record, produce, and disseminate her work in multiple ways suggests a different type of "writing"—a scholarship that surpasses an impression on a sheet of paper to signify the echoes of the jots and scratches on the page. Her recognition as a model performance ethnographer is (as I mentioned earlier) not simply because of her creative methods but also because of the presentation of her work. Her collection of sounds from the American South to the Haiti expedition should stand alone as academic scholarship. The FWP, to which I referred earlier, holds one of the few collections in which Hurston provides a meta-pedagogical explication of the folklore; yet in the Haiti work, which is considered the first exhaustive and descriptive account of Haitian Voudou, the recordings offer limited annotations. The sounds of the ritual performances stand alone, inviting the listener to participate in a dialogue of the seamless stream of chants, dances, and prayers in Haitian Kreyòl. Interestingly, in her printed work Hurston cites very little, if any, from her audio documentation, but rather relies on her written personal engagement with such ritual performances to document her fieldwork. The aural and the written overlap and stand parallel to be sure, but they provide two diverse and distinct academic projects on the topic of Voudou and Haiti. The written Hurston has been privileged in academic corridors, but the aural Hurston is just as valuable and rich and provides a plethora of data not addressed in her printed work.

The digitized Hurston is accessible, if one can find her. Unlike the Jazz Loft Project, most of Hurston's digitized recordings have not undergone exhaustive and comprehensive cataloging and encoding.[28] Perhaps one reason is that her digital sound recordings exist under the auspices of others such as Alan Lomax and Herbert Halpert, two white men. The American Folklife Collection (of the Library of Congress, which houses the sound collection of the Lomax, Hurston, and Barnicle expedition of 1939) lists Hurston as recorder, interviewer, and collector.[29] The project description of the Southern States (WPA) recordings (also in 1939), led by Herbert Halpert, cites Hurston as one of the recorders.[30] Yet the official Library of Congress Folklife Center catalog lists Halpert as the sole recorder and archivist.[31] In one of his final interviews, Stetson Kennedy, who directed the WPA "America Eats" project in the South—which was also under the auspice of Halpert—recalls that because of Jim Crow he dispatched Hurston alone to African American communities to record culinary practices.[32] The writings and photographs of this project have garnered attention, but the vast audio archive from "America Eats" has yet to be exhumed. Hurston's position as female and African American profoundly influences the accessibility and legibility of her work. To find her digital sound work is to go through the work of others. Yet to fully begin to explore the range and significance of Hurston's sound work is also to contemplate the race and gender cultural politics of her time and now, in the twenty-first century. I recall the news in 1997 when unpublished plays and essays by Hurston were "found" in the Margaret Mead audiovisual collection in the Library of Congress.[33] I remember Alice Walker's search for and discovery of Hurston's unmarked grave (a quest that brought greater attention to Hurston's literary work). Yet Hurston's available digitized audio work is placed in the archives of others, some contributions marked, some unmarked. Unlike the sliver of Hurston found in the "America Eats" project, most of her audio remains uncataloged, an omission that limits the knowledge of Hurston's generative work.

TRACK 5: Zora, Digital Sound, Remix

As an artifact, recorded audio produces an aura of authenticity or realness and even a sense of beauty for the listener. In my first encounter with hearing Hurston's voice from a digitized recording, I felt a sense of intimacy, a newness of wonder of what her life must have been like at that time and place. A sense of excitement and eagerness to share the "folklore" exudes from her voice. The imbalance of recording speed along with the extraneous distant and sometimes not-so-distant sounds adds to a sense of awe and mystery surrounding Hurston. Through performance one might consider the objects used in the recording: why that particular recording device (its history, commercial use), how it was used in the exchange with Hurston, and what its interpellations meant for contemporary audiences in their everyday lives. As with the mystery that continues to surround Hurston's personal story (her ability to obscure fact with fiction in documenting her own life—her date of birth, number of marriages, etc.), she brilliantly and strategically used performance as a means of obtaining her desired goals from specific audiences.[34] Performance as a form of analysis at its core interrogates the precept of goal and audience, a position clearly embraced by Hurston. Equally, her digitized voice evokes beauty; her off-toneness and imperfections conjure the traditions of blues music performance. As with most of her work, the recordings disrupt the concept of a standard of beauty for a preoccupation with the haunted space between form and content. Like the sonic mix weaving Hurston's voice at the Red Rooster, the digitized sound recordings allow for more performances of her voice in more venues, with each performance spawning divergent and diverse ways of engaging with her and unearthing more about her. Derrida appends a prospectus on hauntology in *On Hospitality*, in which he makes a compelling case for engagement with the thing (the other).[35] Ultimately, he suggests, sincere engagement will produce an endless stream of discourses, readings, and interpretations. The sound mix the DJ was spinning at the Red Rooster on that winter evening continues to haunt, not only as a confluence between digital and humanistic inquiry but also in its creation of a performative space made possible through technology and sound. In this moment of "re-," consider the multifaceted nature of sound and the enhancement of humanistic inquiry with digital technology, as we recall and remember those whose contributions could be lost, hidden, or unmarked. Patricia Hager recognizes Hurston as museum—a reservoir of folklore, history, a preserver of culture.

NOTES

The chapter opening epigraphs are from Zimmerman, "The Sounds of Zora Neale Hurston's *Their Eyes Were Watching God*"; Hurston, *Dust Tracks on a Road, an Autobiography*, 13; and Hurston and Wall, *Memoirs, and Other Writings*, 804. The epigraphs following the subhead "Track 4" are taken from Brooks, "Sister, Can You Line It Out?," 618, 26.

1. Simone, "Feelin' Good."
2. Brooks "Sister, Can You Line It Out?"
3. Buskirk et al., "The Year in 'Re-.'"
4. Browning, *Infectious Rhythm*.
5. Buskirk et al., "The Year in 'Re-,'" 127.
6. Weheliye, *Phonographies*; Gates, *The Signifying Monkey*.
7. Gates, *The Signifying Monkey*. For polyvocality and orature, see Ngũgĩ wa, *Penpoints, Gunpoints*.
8. Weheliye, *Phonographies*, 114.
9. Phelan, *Unmarked*.
10. Derrida, *Specters of Marx*.
11. Taylor, *Archive and the Repertoire*.
12. Miller [DJ Spooky Kid], *Rhythm Science*; Browning, *Infectious Rhythm*.
13. Miller [DJ Spooky Kid], "That Subliminal."
14. Lee and Gura, *Oral Interpretation*.
15. Jackson, *Professing Performance*.
16. Barthes, *Writing Degree Zero*; Conquergood, "Ethnography, Rhetoric."
17. Two examples from anthropology are Rosaldo, *Culture and Truth*, and Clifford, *Predicament of Culture*.
18. Conquergood, "Ethnography, Rhetoric."
19. Harris, "Derrida Meets Mandela."
20. Davis, *Blues Legacies*; Denzin, *Interpretive Ethnography*.
21. Walker, *In Search of Our Mothers' Gardens*, 91.
22. Davis, *Blues Legacies*, 3.
23. Hurston, *Mules and Men*, 15.
24. Hurston, *Mules and Men*, introduction.
25. Hurston, *Mules and Men*, 18.

26 Lomax, Hurston, et al., "Field Recordings, Vol. 7, Florida."
27 Florida Memory Project, "Florida Memory—Audio—Zora Neale Hurston." In addition to the Library of Congress (accessed January 13, 2018, www.loc.gov/folklife/guides/Hurston.html), the Florida Memory Project has digitized sound recordings of the WPA's work in Florida and can be accessed online: www.floridamemory.com/onlineclassroom/zora_hurston/documents/audio (accessed January 13, 2018). The Alexander Press database also has a reservoir of digitized work of Hurston with Lomax in Florida and Haiti (accessed January 13, 2018, http://alexanderstreet.com).
28 Smith and Stephenson, "Jazz Loft Project."
29 Lomax et al., "Lomax-Hurston-Barnicle Expedition Collection."
30 See "Florida Folklife from the WPA Collections."
31 See Lomax et al., "Lomax-Hurston-Barnicle Expedition Collection."
32 For more information about the audio "America Eats" project, see Nelson and Silva, "'America Eats.'"
33 See Margaret Mead audiovisual collection (accessed January 13, 2018, www.loc.gov/today/pr/1997/97-065.html).
34 Kaplan, Zora Neale Hurston.
35 Derrida and Dufourmantelle, De L'hospitalité.

WORKS CITED

Baldwin, James. Go Tell It on the Mountain. New York: Delta Trade Paperbacks, 2000.

Barthes, Roland. Writing Degree Zero. London: Cape, 1967.

Brooks, Daphne. "Sister, Can You Line It Out? Zora Neale Hurston and the Sound of the Angular Black Womanhood." American Studies 55 (2010): 617–27.

Brown, Gabriel, Rochelle French, John French, Mary Elizabeth Barnicle, Zora Neale Hurston, Alan Lomax, and Archive of Folk Song (U.S.). Out in the Cold Again. Flyright Matchbox SDM 257. 12-inch LP. 1975.

Browning, Barbara. Infectious Rhythm: Metaphors of Contagion and the Spread of African Culture. New York: Routledge, 1998.

Buskirk, Martha, Amelia Jones, and Caroline A. Jones. "The Year in 'Re-.'" Artforum International 52, no. 4 (2013): 127–30.

Clifford, James. The Predicament of Culture: Twentieth-Century Ethnography, Literature, and Art. Cambridge, MA: Harvard University Press, 1988.

Conquergood, Dwight. "Ethnography, Rhetoric, and Performance." Quarterly Journal of Speech 78 (1992): 80–97.

Conquergood, Dwight. "Performing as a Moral Act: Ethical Dimensions of the Ethnography of Performance." Literature in Performance 5, no. 2 (1985): 1–13.

Conquergood, Dwight. "Poetics, Play, Process, and Power: The Performative Turn in Anthropology." *Text and Performance Quarterly* 9 (1989): 18–95.

Conquergood, Dwight. "Rethinking Ethnography: Towards a Critical Cultural Politics." *Communication Monograph* 58 (1991): 179–94.

Cronin, Gloria L. *Critical Essays on Zora Neale Hurston*. New York: Prentice Hall International, 1998.

Davis, Angela Y. *Blues Legacies and Black Feminism: Gertrude "Ma" Rainey, Bessie Smith, and Billie Holiday*. New York: Pantheon Books, 1998.

Denzin, Norman K. *Interpretive Ethnography: Ethnographic Practices for the Twenty-First Century*. Thousand Oaks, CA: Sage Publications, 1997.

Derrida, Jacques. *Specters of Marx: The State of the Debt, the Work of Mourning, and the New International*. New York: Routledge, 1994.

Derrida, Jacques, and Anne Dufourmantelle. *De L'hospitalité*. Paris: Calmann-Lévy, 2005.

Dorst, John Darwin. *The Written Suburb: An American Site, an Ethnographic Dilemma*. Philadelphia: University of Pennsylvania Press, 1989.

Duck, Leigh Anne. "'Rebirth of a Nation': Hurston in Haiti." *Journal of American Folklore* 117, no. 464 (2004): 127–46.

"Florida Folklife from the WPA Collections, 1937–1942." Library of Congress Digital Collections. Accessed November 8, 2017. http://memory.loc.gov/ammem/collections/florida.

Florida Memory Project, Florida Folklife Collection. "Florida Memory—Audio—Zora Neale Hurston." Florida Division of Library and Information Sciences. Accessed January 13, 2018, http://floridamemory.com/audio/hurston.php.

Foucault, Michel, and Colin Gordon. *Power/Knowledge: Selected Interviews and Other Writings, 1972–1977*. New York: Pantheon, 1980.

Frever, Trinna S., and Zora Neale Hurston. "'Mah Story Ends,' or Does It? Orality in Zora Neale Hurston's 'The Eatonville Anthology.'" *Journal of the Short Story in English*, no. 47 (Autumn 2006): 75–86.

Gates, Henry Louis. *The Signifying Monkey: A Theory of Afro-American Literary Criticism*. New York: Oxford University Press, 1988.

Goffman, Erving. *The Presentation of Self in Everyday Life*. Woodstock, NY: Overlook Press, 1973.

Hager, Patricia. Personal interview. Lewiston, May 13, 2014.

Hamilton, Neil A. *Lifetimes: The Great War to the Stock Market Crash: American History through Biography and Primary Documents*. Westport, CT: Greenwood Press, 2002.

Harris, Verne. "Jacques Derrida Meets Nelson Mandela: Archival Ethics at the Endgame." *Archival Science* no. 1 (2011): 113–24.

Howard, Lillie P. *Alice Walker and Zora Neale Hurston: The Common Bond*. Westport, CT: Greenwood Press, 1993.

Hurston, Zora Neale. *The Complete Stories*. New York: HarperCollins, 1995.

Hurston, Zora Neale. *Dust Tracks on a Road, an Autobiography.* Philadelphia: J. B. Lippincott, 1942.

Hurston, Zora Neale. *Mules and Men.* Philadelphia: J. B. Lippincott, 1935.

Hurston, Zora Neale. *Their Eyes Were Watching God: A Novel.* New York: Perennial Library, 1990.

Hurston, Zora Neale, et al. "South Carolina, May 1940, Commandment Keeper Church, Beaufort—Field Footage." 1940. Videocassette (digital Betacam), c. 42 min. Margaret Mead Collection, Library of Congress.

Hurston, Zora Neale, and Carla Kaplan. *Every Tongue Got to Confess: Negro Folk-Tales from the Gulf States.* New York: HarperCollins, 2001.

Hurston, Zora Neale, Carla Kaplan, and Harris Collection of American Poetry and Plays (Brown University). "Poetry 1926–1950." In *Zora Neale Hurston: A Life in Letters,* by Carla Kaplan. New York: Doubleday, 2002.

Hurston, Zora Neale, and Cheryl A. Wall. *Folklore, Memoirs, and Other Writings.* New York: Library of America, 1995.

Jackson, Shannon. *Professing Performance: Theatre in the Academy from Philology to Performativity.* Cambridge, UK: Cambridge University Press, 2004.

Kaplan, Carla. *Zora Neale Hurston: A Life in Letters.* New York: Doubleday, 2002.

Lee, Charlotte I., and Timothy Gura. *Oral Interpretation.* 6th ed. Boston: Houghton Mifflin, 1982.

Lomax, Alan, Zora Neale Hurston, John A. Lomax, et al. "Field Recordings. Vol. 7: Florida, 1935–1936." Sound recordings. Document Records, Vienna, Austria, 1997.

Lomax, Alan, Zora Neale Hurston, Mary Elizabeth Barnicle, et al. "Lomax-Hurston-Barnicle Expedition Collection." 1935. Sound recordings. Archive of American Folk Song, Library of Congress. Accessed November 8, 2017. http://lccn.loc.gov/2008700301.

Miller, Paul D. [DJ Spooky Kid]. "That Subliminal." In *Rhythm Science.* Mediawork Pamphlet Series. Cambridge, MA: Mediawork/MIT Press, 2004.

Moten, Fred. *In the Break: The Aesthetics of the Black Radical Tradition.* Minneapolis: University of Minnesota Press, 2003.

Nelson, Davia, and Nikki Silva [The Kitchen Sisters]. "'America Eats': A Hidden Archive from the 1930s." NPR, November 19, 2004. www.npr.org/2004/11/19/4176589/america-eats-a-hidden-archive-from-the-1930s.

Ngũgĩ wa, Thiong'o. *Penpoints, Gunpoints, and Dreams: Toward a Critical Theory of the Arts and the State in Africa.* New York: Oxford University Press, 1998.

Phelan, Peggy. *Unmarked: The Politics of Performance.* New York: Routledge, 1993.

Plant, Deborah G. *"The Inside Light": New Critical Essays on Zora Neale Hurston.* Santa Barbara, CA: Praeger, 2010.

Rosaldo, Renato. *Culture and Truth: The Remaking of Social Analysis.* Boston: Beacon Press, 1989.

Schechner, Richard. *Performance Theory.* Rev. ed. New York: Routledge, 2003.

Schechner, Richard, and Mady Schuman. *Ritual, Play, and Performance: Readings in the Social Sciences/Theatre*. New York: Seabury Press, 1976.

Simone, Nina. "Feelin' Good." *I Put a Spell on You*, Philips Records, 1965.

Smith, W. Eugene, and Sam Stephenson. "The Jazz Loft Project: Photographs and Tapes of W. Eugene Smith from 821 Sixth Avenue, 1957–1965." Durham, NC: Knopf Center for Documentary Studies, Duke University, 2009.

Taylor, Diana. *The Archive and the Repertoire: Performing Cultural Memory in the Americas*. Durham, NC: Duke University Press, 2003.

Turner, Victor Witter. *From Ritual to Theatre: The Human Seriousness of Play*. New York: PAJ Publications, 1982.

Turner, Victor Witter, and Richard Schechner. *The Anthropology of Performance*. New York: PAJ Publications, 1986.

Walker, Alice. *In Search of Our Mothers' Gardens: Womanist Prose*. San Diego, CA: Harcourt Brace Jovanovich, 1983.

Weheliye, Alexander G. *Phonographies: Grooves in Sonic Afro-Modernity*. Durham, NC: Duke University Press, 2005.

Zimmerman, Katherine Anne. "The Sounds of Zora Neale Hurston's *Their Eyes Were Watching God*: Blues Rhythm, Rhyme, Repetition." MA thesis, University of North Carolina at Greensboro, 2013.

03

RHETORICAL FOLKNESS

Reanimating Walter J. Ong in the
Pursuit of Digital Humanity

JONATHAN W. STONE

I begin this chapter with a confession.

I am a lifelong fan of Phil Collins, the British drummer turned pop superstar.

Since I moved on from grade school—when "Against All Odds" was a radio hit and such an affinity still had social cachet—my fandom is a secret I have been able to conceal with limited success over the years. I keep my public listening contained to headphones, but my neighbors know. They heard me through an open window singing "Sussudio" in the shower, and it is not something I am likely to live down.

Recently, while searching YouTube for interesting Phil Collins tunes, I stumbled on a compelling version of "In the Air Tonight" sung by the choral group Choir! Choir! Choir![1] C!C!C! is an ad hoc ensemble that meets in a bar in Toronto once a week to sing pop music. The group was formed by Daveed Goldman and Nobu Adilman and modeled after an Argentinean peña, which, as Goldman explains, "is just a place where people can go and hang out, even at three o'clock in the morning, and sit at tables with their guitars, drinking red wine or Coca-Cola, and stay up all night singing Salteño folk songs."² Goldman and Adilman's choir operates on a completely volunteer, no-audition basis. Folks just show up and sing. Adilman directs the group and Goldman accompanies on guitar. Together, they arrange the harmonies and make the song selections, which range from A-ha's classic "Take On Me" to Robyn's more contemporary "Dancing on My Own."

In a 2012 NPR story on the group, Adilman relates that the choir started with just twenty participants but quickly grew to a group of over a hundred singers. In the piece, C!C!C! members speak of the ways that the group became a thriving and meaningful community for its participants—a kind of musical refuge—and how the group filled a gap in their social lives. Since its formation and success, Goldman and Adilman now frequently take C!C!C! into the larger Toronto community where they perform in hospitals and for veterans and other groups. In February 2016 the choir raised C$60,000 for Syrian refugees seeking safety in their city. In addition to this community outreach work, C!C!C!'s musical rhetoric has been used in both satire and eulogy. In 2014 they posted a slightly revised version of Sting's "Russians" in response to Vladimir Putin's homophobic comments about LGBTQ athletes participating in the Sochi Winter Olympics, and in 2016 C!C!C! honored the memory of Prince by inviting 1999 people to join them in a stirring rendition of the late singer's "When Doves Cry." Much too large to fit in their regular barroom, this expanded C!C!C! filled Toronto's Massey Hall.

In my initial YouTube encounters with C!C!C!, I was moved by the ways that they projected a rare spirit of participation, care, and celebration of shared cultural experience. That spirit reminded me of the work of another favorite (much more socially acceptable) artist of mine, Pete Seeger. For over seventy years Seeger was an ambassador of vernacular music, an amplifier of marginalized voices, and an untiring advocate for cooperation—for coming together in both song and labor to remember history and plan for the future. C!C!C!'s undertaking resonates with Seeger's values. They are a diverse group and not particularly virtuosic, but when you hear them it is clear they have spent time and care rehearsing the songs in their repertoire. Aside from our shared love for Phil Collins, I was surprised by my strong emotional response to the group's performances. I never cared much for Gordon Lightfoot's "If You Could Read My Mind," for instance, or thought to juxtapose it with "Basket Case" by Green Day, but re-presented in this context, those tunes took on a vibrant quality, a kind of digital vernacular newness. I use the term "vernacular" here in its most basic sense to mean "everyday language," but Houston A. Baker's definition is also useful. He describes vernacular expression in contrast to "high art" as "arts native or peculiar to a particular country or locale."[3] C!C!C! brings together two distinct vernacular practices, two "locales" of shared public experience: popular music and the digital interface. In a strange but unmistakable way, the group's YouTube channel has become a spontaneous digital archive of late twentieth and early twenty-first-century popular culture, which many of us are still ensconced within. More than a playlist on iTunes or a mix CD you might give to a friend, the embodied nature of community singing stored and shared on the public network of YouTube is affecting, drawing me in as a proxy participant. I spent time not just listening to the group but also singing along. In fact, I was inspired to try it out myself. Soon after discovering C!C!C! I brought a guitar and a pile of lyrics sheets to a friend's party and, without much prompting, had twenty people singing "In the Air Tonight" on a porch in Champaign, Illinois. It turns out I'm not the only one with a secret.

Choir! Choir! Choir! and similar groups embody an overlap between popular culture, sound as orality, and the archive that is only beginning to be imagined and activated, let alone theorized, within our contemporary digital culture. The Toronto-based project is a powerful example of what might be called "digital humanity"—a kind of vernacular residuum resulting from the same digital affordances, technologies, and methodologies now being utilized and studied by the emerging institutional formation called digital humanities. Ungoverned by any institution or discipline, digital humanity describes the myriad ways humans are linked together digitally through the common cultural experiences, tools, networks, and technological ambience of the electronic age. C!C!C! brings together the rhythms of popular music with the algorithms of the digital archive and in doing so becomes part of the growing and indelible imprint of digital affordances on human memory-making and -keeping. C!C!C! represents just one of many ways the human coalesces with the digital to preserve, enhance, and perpetuate the rhythms of cultural memory and, by extension, the refrain of human values.

This chapter is my response to the call from the editors of *Provoke!* for essays to better understand digital sound studies by tracing deep histories of digital sound technologies and their predecessors and also to critically evaluate how technology continues to shape auditory culture. To this end, I draw together concepts from ancient and contemporary rhetoric to theorize and historicize this notion of digital humanity for the future of digital sound studies. The rhetorical tradition has been underutilized as a tool for understanding sound in the larger field of sound studies—an oversight I believe is (at least in part) tied to the disavowal of Walter J. Ong in prominent sound scholarship. I will argue that Ong anticipated the current media landscape, including the circumstances I have designated as digital humanity, and that he tied these contemporary mediated realities to both sound studies and rhetoric.[4]

A concept like digital humanity is plausible only because its evidence is everywhere. Everyday lives are becoming more and more reliant on digital tools, not just for connecting and relating to one another, but also in the rhetorical practice of preserving and propagating cultural values and systems of civic belief. Digital humanity is integral to this emergent vernacular digital culture and, as I will argue, can be understood and made useful for sound scholars in terms first conceptualized by ancient rhetoricians, and then retheorized by Ong. Given Ong's fall from scholarly fashion, I will tread carefully if deliberately through that critique in order to articulate a future of digital sound studies that is open to both rhetorical interrogation and a remixed and reanimated Ong. This reanimation, I argue, will help to forge an incipient bridge between sound studies and rhetoric based around a reconceptualization of Ong's sonic theory of secondary orality, one Ong rooted to sound and rhetoric. In other words, Ong's theory reminds us what is rhetorical about sound studies. Rhetoric, generally studied as a tool for persuasion, also has deep connections to sonic ways of value- and knowledge-making. As such the field of sound studies—particularly at the cutting edge of its marriage of the vernacular and the digital—can be better understood and articulated when rhetoric is included as part of the field's conceptual Pro Tools.

Reanimating Walter J. Ong

Once known as a preeminent sound theorist whose "version of the 'great divide' between orality and literacy [for a time] dominated the approach to literacy," Ong now occupies a tenuous position within the fields of literacy studies and sound studies.[5] In fact, Ong has become a kind of Phil Collins figure: both had hits in the 1980s, and while many of us know the words and can sing along to both, it is becoming harder and harder to admit it in public.

Ong was associated with Marshall McLuhan and the influential Toronto School of communication theory.[6] As such, he emphasized key epistemological differences between orality and literacy, arguing for the need to "reawaken the oral [and sounded] character of language" within the scholarly world. He taught that an emphasis on visual, literate (and by extension, logical, empirical, and positivistic) epistemologies led to a diminished understanding of oral/aural types of knowledge. Ong argued that the sound of the voice is an essential feature to understanding humanity and that "the phenomenology of sound enters deeply into human beings' feel for existence, as processed by the spoken word."[7] These claims won the theorist wide acclaim as an innovator and, for a time, helped to bring sound into scholarly vogue. However, critics have since maligned Ong and his contributions as part of a larger grouping of misguided "phonocentrists" who mistake voice and sound for a metaphysical and mythic presence. One of the more notable critics, Jonathan Sterne, describes Ong's position as theocentric and as part of an "audiovisual litany" that seeks to privilege sound over visuality in a kind of hierarchy of the senses. For Sterne, Ong's championing of orality is merely "a restatement of the longstanding spirit/letter distinction in Christian spiritualism."[8] Sterne's perspective is persuasive, and his voice, alongside those of other prominent critics, has contributed to Ong's work falling out of scholarly fashion.[9]

Seeking to "recover" Ong or to rationalize his spiritualism would be futile. He was, after all, *Father* Ong—a Jesuit priest—and he would not have felt compelled to rationalize his spiritual focus either. Instead, in much the same way that C!C!C! brought Phil Collins into a new sonic space, it is more useful to reanimate and redress Ong's intellectual contributions within a secular, rhetorical paradigm, directing attention to the ways he connected his theories of sound to a more technologically diverse understanding of human flourishing—to digital humanity. Ong's work is important not so much for the ways that he tended toward essentialism but in the ways that he understood and began to theorize contemporary society as a hybrid of the traditional and the technological and what that hybridity might have to teach us about human value-making as we move deeper into the electronic age.

THINKING CONJUNCTURALLY: Epideictic Rhetoric, Folkness, and Ong's Secondary Orality

I have already used the term "digital humanity" to gesture toward the ways that people utilize technology to generate new knowledge, tools, and networks for understanding the world and other people. However, the notion that these behaviors lead to the disruption, modification, and even creation of new systems of value has ancient origins. Aristotle conceptualized the deep, humanistic work of belief formation and propagation as a species of rhetoric he called *epideiktikon*, or "epideictic," and he used the word to describe the value-making oratory inherent to ceremonial, ritualistic, and poetic discourse. "Epideictic" remains common parlance in rhetorical criticism, but there are many synonyms across the disciplines. In the 1950s and 60s, for example, musicologist Charles Seeger used the word "folkness" in much the same way. Echoing Aristotle, Seeger defined folkness as the "funded treasury of attitudes, beliefs, and feelings toward life and death, work and play, love, courtship and marriage, health and hearth, children and animals, prosperity and adversity—a veritable code of individual and collective behavior belonging to the people as a whole."[10] While this definition hails from a particularly poignant moment of folk revival, I contend that it points to a more or less universal idea about the ways that humans build systems of value and public memory together in vernacular, or everyday, discursive spaces. Neither epideictic nor folkness is inherently sonic, but both have a close historical relationship with sounded and rhythmic expression, which can also be found commonly at the vernacular level, particularly when paired with rhetorics of remembering. Sound's rhetorical folkness is alive and well within digital culture-making. It is at the heart of CICICI's ethos, for example, but it can also be found in any user-generated or open-source community where memory making and memory keeping have become a public affair due to the increasing ubiquity of electronic affordances: smartphones, inexpensive high-definition sound and video recording devices, and networked platforms such as YouTube. For example, consider "OutKasted Conversations," Regina Bradley's YouTube archive of public conversations about the hip-hop group OutKast and contemporary black southernness. "OutKasted Conversations," addressed in chapter 5 of this book, brings together a number of sonic elements—OutKast's music, digitally mediated conversation, and user-generated comments—all of which contribute to the rich digital humanity of Bradley's work and archive.

In his 2013 collection *The Sound Studies Reader*, Jonathan Sterne asserts that a primary goal for the future of sound studies should be to "think conjuncturally about sound and culture."[11] I have been working here to draw connections, then, between disciplines and terms in order to map these potential conjunctures. Ong also worked conjuncturally, making explicit the connections between value-making, rhythmic sound, memory, and technology in his recurring notion of "orality." Orality was derived from the system of thought known as "media ecology," a central philosophical tenet of the Toronto School.[12] Media ecology's trajectory holds that technological mediation is central to understanding the development of human consciousness and has traversed four major "ages": tribal, literate, print, and electronic. Operating mainly within the trajectory of Western cultural history, Ong's work deals in large part with the transition between each age, and relabels "tribal" with his term "oral." Ong was fascinated by the liminal moments between the ages—with the profound transference that occurs as one dominant mode of communication gives way to the next. We can get a sense for Ong's modus operandi in the title of what is arguably his most intellectually enduring work, *Ramus, Method, and the Decay of Dialogue: From the Art of Discourse to the Art of Reason* (1958). In *Ramus*, Ong chronicles the cultural impact of the printing press, which included a shift from the dominance of speech and dialogue in the public sphere and led, through new emphases on method and "reason," to the cultural circumstances that directly preceded the Enlightenment. For Ong, the historical trajectory that began with preliterate culture and continued through the age of print always includes a steady march away from "aural-type phenomena" and toward ways of knowing structured by "visual-type" methodologies and the abstract thinking made possible by the affordances of literacy and, by extension, technology.[13] This was the case, at least, until the mid-twentieth century and the dawn of a new age.

Ong develops the notion of "secondary orality" as a way of describing the state of human consciousness in the then-emerging electronic age in which the visual's dominance was beginning to wane. The electronic age is one imbued with a "high-technology ambience," where "a new orality is sustained by telephone, radio, television, and other electronic devices that depend for their existence and functioning on writing and print."[14] Ong gives primary and secondary orality as key concepts for understanding the impact and permutations of media ecology's continuum across history. Secondary orality is "new" because technological advance brought sound back into prominence within communication technologies in a way not emphasized since the days of ancient, or "primary," orality.

Ong's writings on secondary orality, however, are somewhat limited. Ong scholar and rhetorician Abigail Lambke argues that "secondary orality should be read as incomplete, suggestive, and germinal" (and this, perhaps, is the right approach to working with Ong in general). Lambke points toward two particularly cogent elements within Ong's cursory beginnings that help define this slippery term.[15] First, Ong writes that "new orality has striking resemblances to the old in its participatory mystique, its fostering of a communal sense, its concentration on the present moment, and even its use of formulas." Next, secondary orality also "promotes spontaneity because through analytical reflection we have decided that spontaneity is a good thing. We plan our happenings carefully to be sure that they are thoroughly spontaneous."[16] Rendered this way, secondary orality reminds us that humanity, as we currently experience it, is a mix of both traditional and progressive paradigms and cannot be otherwise. Digital humanity, then, reveals itself within this symbiosis of the past, present, and future as emerging technologies present opportunities to participate in, preserve, and be "conspicuously spontaneous" in our various technologically enhanced social interactions.

Recall again the ways that Toronto's Choir! Choir! Choir! embodies each of these elements. Participants gather together in planned spontaneity to sing. Their performances are recorded, archived on YouTube, and thereby distributed to the world where we can then participate with them and even emulate them in our own communities. This is in many ways a stunning reversal of the prediction made by Robert Putnam in his popular book *Bowling Alone*, published at the dawning of the twenty-first century. Instead of bowling alleys emptying of people due to technology's "individualizing" tether, people are gathering in public places to perform the ways technology brings them together in common folkness.[17] Indeed, digital networks like YouTube are pulling dispersed individuals into purposive communities and enabling the singular voice of radio, television, and internet to become a collective one.

And YouTube is not the only space where technological advance is creating new communities of participatory sonic culture. The Berlin-based online audio distribution platform SoundCloud, for example, has become a hub for musical collaboration, sample sharing, and new-artist promotion and has a reported forty million registered users and five times that many listeners.[18] Also, smartphone platforms such as Instagram, Vine, and Snapchat, which allow for the quick and simple distribution of vernacular sound and video to a large audience, also meet these criteria (as do video streaming apps like Twitter's Periscope and Facebook Live). With seventy million users (and before its abrupt demise), Vine had become so popular and ubiquitous as to have produced its own "stars"—fascinating evidence of a vibrant community drawing on both traditional (celebrity culture) and emerging paradigms of interaction ("followers," "revines," and "likes"). And while digital platforms and apps are perhaps the most conspicuous places to locate digital humanity, they are not the only places public sonic archives emerge. Libraries and community centers are beginning to invest in open-to-the-public facilities for digital content creation and sharing. The St. Louis public library, for example, recently opened a recording studio—open to anyone with a library card—that has become a popular hub for recording both music and other sonic projects.[19] These various shapes and sounds of digital humanity also represent a realization of Ong's predictions for orality's sweeping communal and participatory potential when sounded through contemporary technologies.

Memory, Archives, and a Step Beyond Secondary Orality

In many ways Ong relies on his audience to intuit a sense of what secondary orality entails by following along carefully with his development of primary orality. Whether primary or secondary, "orality is orality in some ultimate sense," Ong quips with his trademark essentialism, and we are left to assemble the pieces on our own.[20] Ong's interest in the sonic experience of orality is often tied to a deeper human interest in technologies of remembering. And while Ong had an implicit preoccupation with spiritual remembrance (or not forgetting God), we need not be spiritualists to find some insight and value in the importance of memory and its connection to sonic activity. Archives, for example, offer an important site for understanding the negotiation, interplay, and overlap between memory and data, a dichotomy that fits more or less analogously beside the notions of orality and literacy, tradition and progression, folkness and technology—all ideas that I have been engaging here. Orality, if tied to our understanding of the archive, loses much of the acrimony the concept receives from Ong's critics. After all, even before digital or material archives, humans used their memories as archives to preserve important cultural knowledge and also to carefully organize the memorized elements of an eloquent oral performance (a practice that continues).[21]

Ong develops orality's relationship to human memory's potential as an archive along two disparate but related trajectories, one with anthropological ends and the other rhetorical. Each is concerned with what Ong and other media ecologists call "oral-formulaic composition," or the use of rhythmic formulae as a way of preserving memory, knowledge, and culture. Ong jumps to some problematic conclusions in his anthropological research that seem to suggest that literacy develops with clean evolutionary determinism across all cultures, in predictable patterns, and always toward alphabetic literacy.²² This paradigm has come under significant critical scrutiny, the sharpest of which is from ethnographer Brian V. Street, who sees much of Ong's work as methodologically reductive, empirically weak, theoretically deterministic, and based on assumptions about cultures he knew little of.²³ In a like manner, this (pseudo-)anthropological line of thinking does not do much to advance the development of secondary orality.

Within the rhetorical tradition, however, Ong argues that oral formulas as "knowledge storage and retrieval devices" have a rich history.²⁴ Ong connects the orality of ancient rhetorical theory to the secondary orality of his twentieth-century moment. The rhetorical tradition has its roots in ancient poetic traditions in which the formulaic and rhythmic memory aids, fashioned as oral mnemonic devices, passed as oral tradition from the Homeric epoch to into later antiquity. The use and memorization of poetic figures (commonplaces) and the use of carefully curated *topoi* (topics) are well-documented practices in both the teaching and performance of rhetoric in ancient Athens. Writing about the methods of ancient teachers of rhetoric known as the Sophists, George Kennedy relates that even "as the composition of oral poetry and the oratory in it was built up with blocks of memorized material adapted to a variety of situations, so sophistic oratory was to a considerable extent a pastiche, or piecing together of commonplaces, long and short."²⁵ Aristotle catalogues many of these "formulary materials" (as Ong calls them) in his *Rhetoric*, written in the fourth century BCE; he was followed in this practice by Roman orator-teachers Cicero and Quintilian in the first century of the Common Era.

Writing at the end of the 1960s, Ong points to the folk revival—to folkness!—as a site of secondary orality where this same kind of oral-formulaic discourse of public memory reemerges. For Ong, the appeal of folk song "derives from the overwhelming persuasion of its devotees that it is of great antiquity (often it is not) and connects with their past."²⁶ In the United States, folk "revival" in the early and mid-twentieth century revolved around the search for and archiving of vernacular artifacts that reverberated with the cultural memory cataloged in Charles Seeger's earlier definition, but it also engaged with a kind of longing for authenticity. "Folklife" archives have become an important part of countless communities and are housed (often with digital components) in libraries and universities across the United States, with the preeminent example at the Library of Congress. Ong, however, pushes past the idea of folklife as something that should only be engaged within a careful, cataloged archive and moves toward a more dynamic folkness of innovation, satire, improvisation, and play that begins to emerge as part of the rhetorical life cycle of figures and commonplaces—including their eventual decay and/or descent into cultural cliché.[27] Digital sound studies should be similarly postured toward sound's various permutations as shaped by and through digital platforms and tools (which is important work), but sound studies should also seek to understand how those sonic permutations are resonant with and respondent to the dynamics of folkness mentioned above. This orientation is distinctly rhetorical, as sound's radical potential for influence is tied (here rephrasing Aristotle's famous definition of rhetoric) to the perceived uses for and malleability of the figures and commonplaces of shared cultural memory in any given scenario.[28] One brief, humorous example of this might be found on a recent horror-film forum in the open-source sound community freesound.org. Freesound allows users to post sound files and respond to requests for specific sounds and music needs. Foley artist AlienXXX posted a one-second audio file titled "Blood_splat_015b.wav" with the following description: "I had a sound request for a 'bloodsplat.' Created these sounds by throwing small portions of water or wet sponge. Recorded with a Zoom H1."[29] In this instance, and in many others like it on freesound.org, user interactivity and sound productions are respondent to user need and request. In the process of creation mediated through freesound's forums, communities form, disperse, and reform to the ebb and flow (and blood splats) of digital humanity.

Here again, Phil Collins becomes a useful lodestar for further understanding of and engagement with how this process works. Over the last half-century we have circulated at least once or twice through what now appears to be the revolving cycle of secondary orality: from the earnest seeking of authenticity, through satire and irony, to innovation and play, and back again. Other closeted Collins fans will remember that before his solo career took off, he performed as part of the progressive rock group Genesis, whose members were known for their innovative musicianship and frequent political themes. As a solo artist, however, his popularity reached its peak during a brief period of (now cringe-inducing) earnestness during the 1980s. Since then, however, Collins's music has remained in the popular sphere, in karaoke bars and among community sing groups like CICICI, to be sure, but also as samples in the work of hip-hop DJs and MCs. In fact, my favorite song, "In the Air Tonight," has been sampled by DMX ("I Can Feel It"), Lil' Kim ("In the Air Tonite"), Nas ("One Mic"), and even the legendary 2Pac ("Staring through My Rearview").[30] Collins's work takes on new life as a DJ's sample. When juxtaposed with hip-hop lyrics and themes, the song functions as a common cultural touch-point—a backbeat—useful in response to (and even subtle commentary on) evolving exigent issues. "In the Air Tonight" carries the broad cultural marker (or commonplace) of emotional intensity, which can be taken up, reworked, and deployed in the praise and blame—the *epideictic* critique—of shared values within and across U.S. popular culture. 2Pac's lyrics, "I wonder when the world stopped caring last night / Two kids shot while the whole block staring," rapped over the iconic keyboard and drums of "In the Air Tonight," are indisputable as poignant oratory and an example of what contemporary epideictic rhetoric sounds like. Collins's work, then, is part of a revolving cycle of rhetorical folkness: from innovative art to tired cliché and back to art—but in new keys and accompanied by new voices.

The folkness of digital humanity, which exists, perhaps, as a step *beyond* secondary orality, takes advantage of a technologically hybrid culture where knowledge/retrieval systems (or "external memory" as we are now wont to call it) have become ubiquitous. In other words, digital humanity is evident in the kind of technical literacy and rhetorical fluency central to the DJ's expertise (mash-up/remix) and can be observed across media and in digital discourse of the everyday. This notion of digital humanity invites a new and emergent folkness that embraces, circulates, and rearticulates each of these stages *ad infinitum*, forever blurring the lines between tradition and progression. Harkening back to both Aristotle's and Lambke's insights, while the ever-changing folkness of digital humanity creates unprecedented opportunities for participation and spontaneity—from open-source software builds to open-audition community choirs—this radical openness also requires new ways of understanding the dissonance of this potential cacophony of competing voices and values. Here rhetoric's concepts and theories, starting with epideictic and blossoming outward, can provide both perspective and conciliatory resonance to these issues as well as those within conversations around digital sound studies more broadly. As mentioned earlier, Ong reminds us what is rhetorical about sound studies. His anticipation of these technologically imbued circumstances and phenomena, his use of rhetoric and rhetorical history to understand and situate them, and his open notion of secondary orality as the mode in which they can be theorized are ample justification for a reconsideration and reanimation not only of Ong's work in relation to digital sound studies scholarship but also of rhetorical studies more broadly.[31]

Consider, in conclusion, how Ong's view on technology (which always exists as a demonstration of the hybridity of oral and literate ways of thinking) speaks to this justification:

> Technologies are not mere exterior aids but also interior transformations of consciousness, and never more than when they affect the word. Such transformations can be uplifting. Writing heightens consciousness. Alienation from a natural milieu can be good for us and indeed is in many ways essential for full human life. To live and to understand fully, we need not only proximity but also distance. This writing provides for consciousness as nothing else does.
>
> Technologies are artificial, but—paradox again—artificiality is natural to human being. Technology, properly interiorized, does not degrade human life but on the contrary enhances it. The modern orchestra, for example, is the result of high technology.[32]

Phenomena such as a C!C!C!, freesound.org, and the DJ sample (and, for that matter, Phil Collins and Walter Ong themselves) exist along a continuum of mediated experience that includes activities that look and sound like Ong's descriptions of primary and secondary orality. I do not need to subscribe to Ong's spiritual ideals to find something transcendent and human about the activities implied by these terms, their various permutations, and the ways that they relate across that continuum. On the other hand, subscribing to and expanding on Ong's frequent use of "rhetoric" to account for the complexity of oral and aural experience has immense potential. This chapter has been about drawing that potential out, connecting rhetorical terms to sonic experiences, and beginning to theorize the folkness of digital humanity.

As we look toward the future of digital sound studies, each of the above frameworks, from secondary orality to digital humanity, usefully conceptualizes the various ways that contemporary vernacular culture is embedded within, performed through, and transformed by digital technology. For those in sound studies, an orientation that acknowledges the inherent folkness of those technologies is, as I have sought to show, a rhetorical orientation. As such, this rhetorical folkness is resonant within digital technologies as the coalescing rhythms and algorithms of past and present memory/value systems resounding in and beyond the code. Within these systems (archives and revolving traditions), groups like Choir! Choir! Choir!, platforms like YouTube, and samples by Phil Collins can be understood not only as sites for the careful study of sound's digital potentials but also as the raw materials available for the crafting of new and rich digital rhetorics.

NOTES

1. Following the group's lead, I generally refer to Choir! Choir! Choir! as C!C!C! throughout for brevity. Find them on their website (accessed January 13, 2018, www.choirchoirchoir.com); on their YouTube channel (accessed January 13, 2018, www.youtube.com/user/CHOIRx3); and on SoundCloud (accessed January 13, 2018, https://soundcloud.com/choir-choir-choir).
2. Myers, "Choirstarters."
3. Baker, Blues, Ideology, 2.
4. Like sound studies, "rhetorical studies" (and, more broadly, the "rhetorical tradition") designates a large and not entirely coherent grouping of scholars and scholarship. This grouping includes at least two prominent disciplinary iterations in the academy: one often found in communication departments, where the scholarly focus tends to be on speech; and another in English departments, where writing and composition are the primary objects/activities of rhetorical inquiry (though this is a somewhat vapid simplification). My evocation of "rhetoric" is meant to name a common tradition that transcends disciplinary divisions. Ong himself studied and wrote of rhetoric and its histories outside of these paradigms and as a professor of literature.
5. Street, "Critical Look at Ong," 153.
6. The Toronto School came about through Harold Innis's and Marshall McLuhan's application and expansion of the work and theories of Eric A. Havelock; all three men were associated with the University of Toronto. This work, which often "explored different implications of ancient Greek literacy to support [its] theoretical approach," was generally focused on the notion that human communication is central to understanding the structures of both human culture and human minds (see Kerckhove, "McLuhan and the Toronto School," 73). Central works include Innis's *Empire and Communications* and McLuhan's *Understanding Media*. Havelock's *Preface to Plato* is also notable here due to its explicit focus on rhetoric.
7. Ong, *Orality and Literacy*, 6, 76.
8. Sterne, *Audible Past*, 16.
9. See Derrida, *Margins of Philosophy*; Street, "Critical Look at Walter Ong."
10. Seeger, "Folkness of the Non-Folk," 3.
11. Sterne, *Sound Studies Reader*, 3.
12. The theory and practice of media ecology came about through the work of McLuhan, Neil Postman, Ong, and other members of the Toronto School. Central works include McLuhan's *Understanding Media* and Postman's well-known (and often critiqued) *Amusing Ourselves to Death*. Media ecology as a theory continues to have a strong academic presence in several anthologies. See, for example,

Crowley and Heyer's *Communication in History* and the journal *Explorations in Media Ecology*, which is devoted to its history and development.

13 Ong, *Ramus*, 107.
14 Ong, *Orality and Literacy*, 11.
15 Lambke, "Refining Secondary Orality," 203.
16 Ong, *Orality and Literacy*, 136, 137.
17 Putnam, *Bowling Alone*, 216–17.
18 See Graham, "Who's Listening to SoundCloud?"
19 See Clark, "St. Louis Central Library."
20 Ong, *Rhetoric, Romance, and Technology*, 284.
21 In addition to their memories, humans also use their bodies as archives. See Taylor, *Archive and Repertoire*, and Schneider, *Performing Remains*.
22 Ong was intrigued by the literary-turned-anthropological work of Milman Parry and his student Albert Lord. Parry is known for his pioneering work in Homeric oral poetry in which he demonstrates convincingly the formulary nature of the *Iliad* and *Odyssey*, which, though eventually written down and deemed "literature," hailed from a much earlier oral tradition. Lord took Parry's work into the former Yugoslavia, where he studied Yugoslav narrative poets who could not read and found the same kinds of formulaic devices at work there that Parry had found in Homer. See Ong, *Orality and Literacy*, 59.
23 See Street, "Critical Look at Walter Ong."
24 Ong, *Rhetoric, Romance*, 285.
25 Kennedy, *Classical Rhetoric*, 28.
26 Ong, *Rhetoric, Romance*, 285.
27 Ong points humorously to the then-contemporary duo Simon and Garfunkel, whose music, he argues, is rife with play on "worn rhetorical clichés." The lyrics offer blatant informality within formal musical settings, "total irony," and "total casualness"—all as playful innovations replacing tired formulaic commonplaces. Recall, for example, the comically mundane line "Citizens for Boysenberry Jam" from their 1968 song "Punky's Dilemma."
28 Aristotle defines rhetoric as "the faculty of discovering in any particular case all of the available means of persuasion." *On Rhetoric* 1.2.1.
29 Freesound.org (accessed January 13, 2018, http://freesound.org/people/AlienXXX/sounds/198827).
30 The site whosampled.com (accessed January 13, 2018) helped me discover this information. According to the site's search engine, "In the Air Tonight" has been sampled in forty-three hip-hop songs to date.
31 For recent work that takes up sound's relationship to rhetoric and the Western rhetorical tradition, see Walker's *Rhetoric and Poetics*, Hawhee's *Bodily Arts*, Johnstone's *Listening to the Logos*, and Rickert's *Ambient Rhetoric*.
32 Ong, *Orality and Literacy*, 82–83.

WORKS CITED

Aristotle. *On Rhetoric: A Theory of Civic Discourse.* Translated by George A. Kennedy. New York: Oxford University Press, 2007.

Baker, Houston A. *Blues, Ideology, and Afro-American Literature: A Vernacular Theory.* Chicago: University of Chicago Press, 1984.

Clark, Patrick. "St. Louis Central Library Recording Studio Has Become a Hotspot for Artist." *Fox 2 Now*, August 11, 2015. http://fox2now.com/2015/08/11/st-louis-central-library-recording-studio-has-become-a-hotspot-for-artist.

Crowley, David, and Paul Heyer, eds. *Communication in History: Technology, Culture, Society.* New York: Routledge, 2011.

Kerckhove, Derrick de. "McLuhan and the Toronto School of Communication." *Canadian Journal of Communication* 14, no. 4 (1989): 73–79.

Derrida, Jacques. *Margins of Philosophy.* Chicago: University of Chicago Press, 1982.

Graham, Jefferson. "Who's Listening to SoundCloud? 200 Million." *USA Today*, July 17, 2013. www.usatoday.com/story/tech/columnist/talkingtech/2013/07/17/whos-listening-to-soundcloud-200-million/2521363.

Havelock, Eric A. *Preface to Plato.* Cambridge, MA: Harvard University Press, 1963.

Hawhee, Debra. *Bodily Arts: Rhetoric and Athletics in Ancient Greece.* Austin: University of Texas Press, 2004.

Innis, Harold. *Empire and Communications.* Oxford: Clarendon Press, 1950.

Jarratt, Susan C. *Rereading the Sophists: Classical Rhetoric Refigured.* Carbondale: Southern Illinois University Press, 1991.

Johnstone, Christopher Lyle. *Listening to the Logos: Speech and the Coming of Wisdom in Ancient Greece.* Columbia: University of South Carolina Press, 2009.

Kennedy, George A. *Classical Rhetoric and Its Christian and Secular Tradition from Ancient to Modern Times.* Chapel Hill: University of North Carolina Press, 1980.

Kerckhove, Derrick de. "McLuhan and the Toronto School of Communication." *Canadian Journal of Communication* 14, no. 4 (1989): 73–79.

Lambke, Abigail. "Refining Secondary Orality: Articulating What Is Felt, Explaining What Is Implied." *Explorations in Media Ecology* 11, nos. 3–4 (December 2012): 201–17.

McLuhan, Marshall. *Understanding Media: The Extensions of Man.* New York: McGraw-Hill, 1964.

Myers, Paul. "Choirstarters: Toronto's Choir! Choir! Choir! Sings the Praises of a (Social-Media-Assisted) Real-Word Experience." Co.Create. *Fast Company.* Accessed April 27, 2015. www.fastcocreate.com/3024227/choirstarters-torontos-choir-choir-choir-sings-the-praises-of-a-social-media-assisted-real-w.

NPR Staff. "In Toronto, an Ad-Hoc Choir Becomes a Community." *NPR Music.*

Accessed April 27, 2015. www.npr.org/2012/12/23/167882491/in-toronto-an-ad-hoc-choir-becomes-a-community.

Ong, Walter J. *Orality and Literacy: The Technologizing of the Word*. New York: Routledge, 1982.

Ong, Walter J. *Ramus, Method, and the Decay of Dialogue: From the Art of Discourse to the Art of Reason*. Cambridge, MA: Harvard University Press, 1958.

Ong, Walter J. *Rhetoric, Romance, and Technology; Studies in the Interaction of Expression and Culture*. Ithaca, NY: Cornell University Press, 1971.

Plato. Dialogues. References to Plato are given in the usual Stephanus numbers.

Postman, Neil. *Amusing Ourselves to Death: Public Discourse in the Age of Show Business*. New York: Penguin, 1986.

Putnam, Robert D. *Bowling Alone: The Collapse and Revival of American Community*. New York: Simon & Schuster, 2000.

Rickert, Thomas J. *Ambient Rhetoric: The Attunements of Rhetorical Being*. Pittsburgh: University of Pittsburgh Press, 2013.

Schneider, Rebecca. *Performing Remains: Art and War in Times of Theatrical Reenactment*. London: Routledge, 2011.

Seeger, Charles. "The Folkness of the Non-Folk vs. the Non-Folkness of the Folk." In *Folklore and Society: Essays in Honor of Benj. A Botkin*. Hatboro, PA: Folklore Associates, 1966.

Sterne, Jonathan. *The Audible Past*. Durham, NC: Duke University Press, 2003.

Sterne, Jonathan. *The Sound Studies Reader*. New York: Routledge, 2012.

Street, Brian V. "A Critical Look at Walter Ong and the 'Great Divide.'" In *Social Literacies: Critical Approaches to Literacy in Development, Ethnography, and Education*, edited by Brian V. Street, 153–59. New York: Routledge, 1995.

Taylor, Diana. *The Archive and the Repertoire: Performing Cultural Memory in the Americas*. Durham, NC: Duke University Press, 2003.

Walker, Jeffrey. *Rhetoric and Poetics in Antiquity*. New York: Oxford University Press, 2000.

II | DIGITAL COMMUNITIES

04

THE PLEASURE (IS) PRINCIPLE

Sounding Out! and the Digitizing of Community

AARON TRAMMELL, JENNIFER LYNN STOEVER,
AND LIANA SILVA

Over the past five years, the *Sounding Out!* Editorial Collective has often heard our sound studies blog *Sounding Out!* referred to as a "labor of love" by our closest colleagues. Usually delivered in a tone that indicates both gratitude and pity—and often preceded by a sigh—the phrase "labor of love" indicates our willingness to "waste" precious uncompensated time from the tenure clock, dissertation timeline, and/or salaried workweek on a blog, with all of the self-indulgence that title entails.[1] Blogging is considered "scholarship lite" among some academic bloggers and tenure-and-promotion committees, who often shunt it directly to the undervalued and much-maligned category of "service."[2] Much like a dysfunctional relationship, our "love" of the field of sound studies (and *Sounding Out!*'s digital medium) has seemingly made us far too willing to donate some serious high-quality, low-value labor on its behalf. (digital) sound studies, we just can't quit you.

    Suckers, right?

    Nope. As quiet as it is kept—and as challenging as shoehorning that labor into already jam-packed, demanding schedules has been—*Sounding Out!* has remained, first and foremost, a labor of *pleasure*. We not only love working on *Sounding Out!*, but it also *feels* good and it is *fun* (two affects rarely mentioned in connection with academic work, particularly in current working conditions). Please do not tell our provosts, deans, chairs, advisors, and/or bosses, because pleasurable labor remains labor nonetheless.

While the massive amounts of fun we actually have while writing, building, curating, editing, representing, designing, tweeting, and so on may come as a surprise to *Sounding Out!* initiates, we'd like to think that our careful readers already sense our enjoyment; that, along with circulating information critical to an ever-increasing fold of sound studies scholars, we have successfully used the digital medium to communicate the very gratifying pleasure we take not only in hosting the "mothership" site and its social media constellation, but in the act of community building itself. In fact, we dare to contend that people who identify as members of the sound studies community also find the persistent, multimodal, participatory, and self-consciously accessible sound studies community *Sounding Out!* has cultivated since 2009 to be a very distinct pleasure.

Despite the pleasure that *Sounding Out!* provokes in authors and readers alike, we nonetheless feel like outsiders in conversations about digital scholarship in the digital humanities. Because many bloggers like us use a digital platform created by someone else, the question of whether blogging really constitutes "making"—a key but contested tenet of digital humanities—is a roiling debate. Of course, as this essay argues, we definitely think it does. Most recently, Debbie Chachra's "Why I am Not a Maker" argues against a strictly defined culture of coding-as-making in the digital humanities, maintaining that it is an oppressive "way of accruing to oneself the gendered capitalist benefits of being a person who makes products." We're stuck in the middle—not quite cool enough to hang with the computer dudes making robots and literature databases, yet somehow also complicit with the gnarly benefits of capitalist production. Our position as outsiders is far from unique; it carries with it the same racial and gendered biases that permeate all domestic spaces of society. Our work—editing, community building, and care—is the undercompensated affective and domestic work of the academy. As bloggers we both *are* and *are not* makers, and therefore we are outsiders.

In this essay, the *Sounding Out!* Editorial Collective explores the central role of carefully tended affect in building a cohesive digital community. We believe that even in terms of intellectual connection, "feeling is first," to quote e. e. cummings.[3] Therefore, we have peppered throughout this essay screenshots from an October 31, 2014, online editorial at Google Hangout as a performative insight into our affect—as individuals and as a collective—

Liana M. Silva
if it weren't for SO! i don't think folks would take me seriously as an intellectual, despite what i write and my interests.

Aaron Trammell
wow.

Jennifer Stoever
and that is why "gatekeeper" doesn't describe us

Aaron Trammell
YES.
We're a causeway
A secret entrance
An underground railroad.
A hack

Liana M. Silva
i should specify that academics wouldn't take me seriously

Aaron Trammell
We're a hack.

Jennifer Stoever
and we are trying to show all the good old boy bullshit the door

Aaron Trammell
An exploit in the system
YES!

FIGURE 4.1 Discussion of our relationship to the social capital of *Sounding Out!*

that functions as an ongoing methodological, sonic, and affective counternarrative within the space of the formal academic essay.[4] Intentionally disruptive, these screenshots provide intimate insights into our editorial process so readers can feel the defensiveness, criticism, and pressure we face on a constant basis; they respond, with boldness and candor, to the feedback we have all encountered throughout our careers about the worthiness of *Sounding Out!* and the blog format. Though *SO!* has become a staple in the sound studies community, we can't help but feel like outsiders looking in to conversations in the digital humanities, which are often centered around the grant-winning merit of boutique digital platforms as opposed to the populous, intimate, and perhaps now antiquated form of web-logging (blogging). But we also want to invite readers new to *SO!* to understand how together we weave a protective sonic web of humor, backtalk, and so-called colloquial language that not only assuages that pressure but also provides an ongoing source of freedom and pleasure, what Sebastian Ferrada calls "an audible badge, a marker of experience rather than a punchline" that constructs "an alternative aesthetic" through speech, accent, and tone. We hope that the selected screenshots provide necessary push-back to the content they interrupt while better contextualizing the love, labor, and passion that have always pushed our humble blog forward.

Combining our frank, spirited, nuts-and-bolts discussions of *Sounding Out!*'s editorial decisions and history, then, with a theorization of digital community and a qualitative analysis of an *SO!* community-member survey, we argue that *Sounding Out!* has only established itself as a trusted and noteworthy venue for sound studies scholarship through an artisan-like approach to community building that fosters an important (yet often missing) *feeling* of community within and without brick-and-mortar institutions. The digital medium in particular facilitates many of the microinterventions *Sounding Out!* stages in the areas of editing, social media engagement, branding, and active readership.

Where It Started

We founded *Sounding Out! The Sound Studies Blog* in 2009 as a way for three academics interested in talking about sound to stay intellectually engaged while physically separated. Little did we know when we first created our WordPress site that seven years later, our project would become, as Jonathan Sterne describes in the survey conducted for this essay, "an interdisciplinary resource for a massive interdisciplinary sound studies community . . . more important than any journal in terms of disseminating ideas and scholarship." Although the respect and trust we have earned from our colleagues has always meant more than official seals of approval, those seals do represent our rapid growth. In 2013 *Sounding Out!* received an ISSN number (2333-0309) from the Library of Congress and in 2014 became one of only ten scholarly sites whose articles the Modern Language Association indexed in its International Bibliography.[5]

In 2007–8 (the year the *Sounding Out!* team came together like Voltron in Binghamton, New York) sound studies as a field remained fairly diffuse and underground. Interest in sound and audio culture seemed constantly emerging and never fully emergent, arising as it did from unique concerns

Jennifer Stoever
so here's my thing--I do like a decentered sound studies from the position of scholarship. being a "side piece" has produced some great work because of the tension between disciplinary location and interdisciplinary inquiry.. but its lonely in the day to day. and even though it seems everywhere right now, how to successfully reproduce another generation of scholars if there are not dedicated grants, only scattered organizations, no departments/programs and most importantly, no jobs. what happens after this hipness wave passes? also, labor-wise, does being a side hustle just extract more labor from the university's side?

Liana M. Silva
yes, those are all very valid points. i like where you're going with this.

Jennifer Stoever
is sound studies just value added?

Aaron Trammell
or worse?

Jennifer Stoever
uh oh

Liana M. Silva
also, how many sound studies scholars can afford to do sound studies, right?

Aaron Trammell
A contition that necessitates vale added

Jennifer Stoever
are we enablers
#thedarksideofSO

FIGURE 4.2 Discussing how *Sounding Out!*'s creation both breaks through the lonely echo chamber faced by most sound studies scholars and creates new—and largely uncompensated—"value" for the neoliberal educational complex.

at different moments in a wide spectrum of academic disciplines—in particular, acoustic ecology, cinema and television, history, anthropology, literature, art history, and ethnomusicology—as well as in thoroughly interdisciplinary fields such as African American studies, American studies, science and technology studies, radio studies, and urban studies. In what Jim Drobnik declared a "sonic turn," a buzz began to circulate around a small canon of recognizable names who published exciting but disparate-seeming monographs.[6] Through Google searches, word of mouth, third-generation photocopies of syllabi, qualifying exam lists, the occasional conference panel, groundbreaking seminars (such as Josh Kun's at the University of California, Riverside, in 2000 and Karen Pinkus's at the University of Southern California in spring 2004), patient, repeated answers to the "What is sound studies?" question from determined graduate students, and dissertations such as Jennifer's in 2007, "sound studies" stubbornly accreted a methodological center.[7]

When Jennifer arrived at Binghamton University as an assistant professor in the fall of 2007, she felt lonely and disconnected from her tight-knit University of Southern California American Studies community and USC's dynamic sound studies nucleus, then composed of Fred Moten, Josh Kun, Bruce Smith, and Joanna Demers. The experience of isolation remains all too familiar for many sound studies scholars even now. There are few, if any, academic job listings for "sound studies" in the United States—and even though positions naming sound studies as a field of interest are becoming more common, they remain in the realm of "a handful." Most academic researchers who work in sound studies are technically hired to do "something else," and interest in sound is presented as a unique methodological take and/or a quirky bonus field. In our current corporate academic speak, it "adds value" to an already solid research profile—which means that, institutionally, sound studies graduate students and professors largely find themselves alone in an echo chamber.

To remedy the sense of stagnation that comes so quickly on the heels of isolation, Jennifer and Aaron began constructing a group called the Binghamton University Sound Studies Collective (BU SSC) as a face-to-face interdisciplinary group to suss out colleagues with even remotely similar interests. At the very least group members had the desire to discuss the exciting new questions surrounding the cultural meaning of sound and listening, seemingly vibrating from everywhere at once. While the group had one well-attended first meeting, a sweet logo, and one hell of a speaker series in 2008–9 (Martin Daughtry, Fred Moten, Frances Aparicio, and Trevor

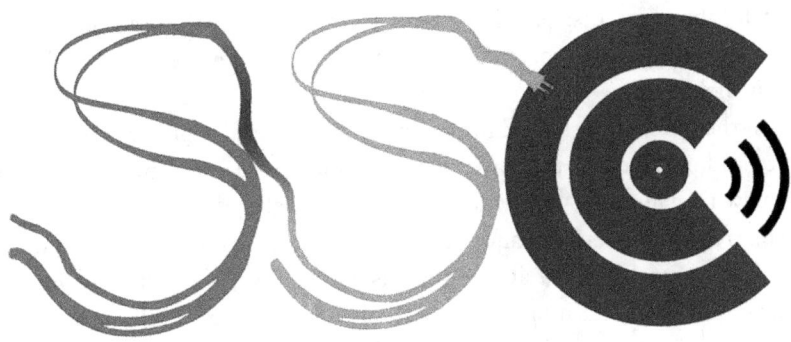

FIGURE 4.3 The official logo of the Binghamton University Sound Studies Collective, designed by Conrad Weykamp, 2011.

FIGURE 4.4 A page from Jennifer Stoever's 2009 daybook, showing our initial planning meeting. We've been a "blog" from day one. IMAGE BY JS.

Pinch), the group dissipated fairly quickly into a lonely listserv and a hardcore handful of awesome grad students who were interested but brand new to the field.

At the time, BU lacked a campus culture and interdisciplinary infrastructure and, apparently, there weren't many interested colleagues. Although somewhat daunted—who wouldn't want to talk about sound while cashing in on free food?—Jennifer and the hardest core of them all, Liana and Aaron, decided to reach beyond BU's highly disciplinary walls and create a virtual community to sustain ourselves as the band broke up. While Jennifer remained at Binghamton, Aaron finished his MA and left for a Rutgers PhD program in media studies; Liana took off to dissertate in Kansas City. But like a CD stuck on repeat, we needed to keep spinning our ideas around to each other. Often. We also hoped that if we put out a virtual bat signal via a blog, we could bring in the folks we were meeting at conferences and reading and writing about via stray journal articles. And they might tell two friends. And so on, and so on. And so on.

The Premise

By design, therefore, we founded *Sounding Out!* as an intervention regarding the notion of affective community as format, logistics, and politics in the field of sound studies. When we say "community," we borrow from Raymond Williams's definition: it reflects "the quality of holding something in common . . . a sense of common identity and characteristics." Interestingly, Williams points out that after the nineteenth century, "community was the word normally chosen for experiments in an alternative kind of group living."[8] Considering that *Sounding Out!* is a space for sound studies aficionados invested in the field in some way—and who are seeking an alternative from silo-bound campus culture—Williams's definition of community as group experiment is fitting.

In addition to Williams's definition, we take inspiration from Jack Bratich's reworking of the term "digital" in "digital community." The predominant understanding of a digital community remains focused on emerging modes of interaction enabled by innovations in computing technology: content management software, Wikis, social media, open-source software, and even MOOCs have been both celebrated and critiqued as new spaces of discourse with the potential to shake things up a bit. We find this definition reductive in its scope, however, as it instantiates the digital as a mode of interaction informed primarily by the materiality of the platform that hosts the interaction. In other words, it is digital because its mode of publication is digital.

But the digital, as Bratich argues, invokes the former definition alongside a second, older connotation: digits as fingers. This understanding of the digital foregrounds moments of craft production and invisible infrastructural labor, as opposed to a definition that focuses instead on the ways being digital often invokes a discussion of platform affordances.[9] When we founded *Sounding Out!*, the blog as a format was swiftly becoming an anachronism of the aughts. We began at a time when the blog no longer was being taken seriously by the mass media—treated instead as a mechanism for instant celebrity or a narcissistic hobby. Despite (or perhaps because of) this, we encountered an intuitive, reliable, and affordable content management system that WordPress had spent the past decade developing (and has spent the time since simplifying to the point of incomprehensibility). So we began the blog with the ethic of a craft circle, trading tips with one another as we learned the WordPress platform. This ethic even seeped into our editorial practices—in which we curate, edit, array, and host with a care often taken only by small, artisanal presses—and circulated through the social media networks of like-minded crafters interested in continuing the dialogue. Jenny Sundén calls this a "transdigital affect," or "a type of corporeal relationality that arises in contemporary passionate encounters with the analog made possible by, or realized through, the digital."[10] *Sounding Out!* uses a digital platform to respond to the traditionalist model of the humanities the way that punk zines allowed radical new voices into the sphere of rock journalism. We are digital activism.

First and foremost, our move to combine craft production with a group experiment in digital community building came from a desire to push the rhetorical boundaries of sound studies and the sensory nature of "writing" itself. We considered, like Mark Sample, "Why must writing, especially writing that captures critical thinking, be composed of words? Why not images? Why not sound? Why not objects? The word *text*, after all, derives from the Latin *textus*, meaning 'that which is woven,' strands of different material intertwined together."[11] The epistemologies through which we apprehend our knowledge affect the modes in which we approach and understand it. Simply put, a sound translated into text is qualitatively different from a live experience of it, and this commonsense fact deserved more than just a nod within our tradition of scholarship.

Working in a "born-digital" format enabled us to think critically about how to present what Marcus Boon calls "sonic realness" in sound studies scholarship and to do it in public, where both our successes and our shortcomings could enable others' work.[12] For example, in addition to embedding sound within posts—with varying degrees of integration—*Sounding Out!*'s monthly podcast has been an important speculative solution to the problem of scholarship through sound. By offering a monthly broadcast with minimal written notes, we have hoped to provoke sound studies scholarship to listen more closely to itself. The podcast space is deliberately unstructured, and broadcasts vary from radio style exposé to interview to digital sound art installation. By remaining freeform, we hope to represent the diverse array of modalities interdisciplinary engagement takes. We serve our constituents by allowing our podcasts to take the forms most necessitated by members of the community.

In addition to the logistics of rethinking the nature of work in sound studies, there has also been an infrastructural need for a communications network. Sound studies in the United States has remained dispersed within the disciplines, even after the European Sound Studies Association formed in 2012. Until 2013 there were no large-scale U.S.-based academic "sound studies" events, although chartered groups represent and vivify the field in several major organizations.[13] Without formal institutionalization in the United States, the field has remained productively critical and refreshingly rhizomic, but its lack of formality has its drawbacks; the exciting interstices of our field remain "dark matter," comprising the bulk of "sound studies" but remaining hidden save for the occasional special-issue spectacular. (Thank you, *Social Text* [2010]! *Performance Research* [2010]! *American Quarterly* [2011]! *differences* [2011]! *Radical History Review* [2015]!)[14] And although the infrastructural work that occurs behind the scenes at conferences and departments across academia is valiant, to say the least, we saw that the field needed a forward-thinking forum that allowed for the expression of its radical sonic epistemologies and interdisciplinary experimentation.

Sounding Out! makes the "interdisciplinary" aspect of sound studies more audible, consistent, and apparent. It highlights existing affinities and makes new contacts between formal groups and individuals by circulating calls for papers on Facebook and Twitter, posting conference previews that address the "state of the field" and cull panels of interest, cross-posting and cosponsoring topical series with groups such as IASPM and *Antenna*, hosting a monthly "Comment Klatsch" open forum (2013–14), and adding media scholar and longtime supporter Neil Verma to the team as SCMS/ASA special editor in 2014. (Neil coordinated guest editors and writers from these organizations.) Very deliberately and through multiple means, *Sounding Out!* spins a center of gravity for sound studies, enabling a sense of community effort, pleasure, and enthusiasm to fuel the push to new areas. Moreover, as we connect with digital humanities scholars via Twitter and HASTAC, we see others asking similar questions about media, format, and research "tools." *Sounding Out!* articulates a #dhsound relationship, even when as "bloggers" we often had felt left out of the DH conversation.

The Politics That Guide Us

In terms of politics, *Sounding Out!* pushes the field through its editorial focus and demography. Every post hosted by *Sounding Out!* provokes conversation about social difference and power, fundamental topics lost or outright evaded as sound studies' newest efflorescence gained momentum in the 2010s. Even as late as January 2015, a sound studies colleague sent out a Facebook message that appeared in Jennifer and Aaron's feeds describing an application received for a new sound studies book series in which the editorial board and prospective authors were all males whose proposed topics blithely ignored the multiethnic and transnational issues at stake in the field.[15] At *Sounding Out!* we proactively think about gender, about race, about class, and about sexuality. By taking an unequivocal stance that politics matter both within and without the field, *Sounding Out!* fosters a material sense for its readers and writers of *being listened to* and *having a voice*, enacting a self-aware and critical public conversation that remains grounded in sound studies' social impact and that continually centralizes the work of scholars who might otherwise be marginalized, even in the generally friendly atmosphere of an emergent field.

Moreover, we don't just talk the politics, we show and prove our commitment to amplify different voices and to reach out to a wider readership. We polish our writing to make it readable: we aim to attract interest rather than assuming it (as much scholarly writing does, to its detriment) and aim for an accessible tone that opens up the rigor of our field beyond the academy. We often describe *Sounding Out!* as the site where our nonacademic friends, family, and colleagues can finally "get" what we have been spending years of our lives studying and see why it matters. At the same time, the *Sounding Out!* Editorial Collective actively recruits an ever-expanding team of regular and guest writers who more accurately represent the demographics of sound studies.

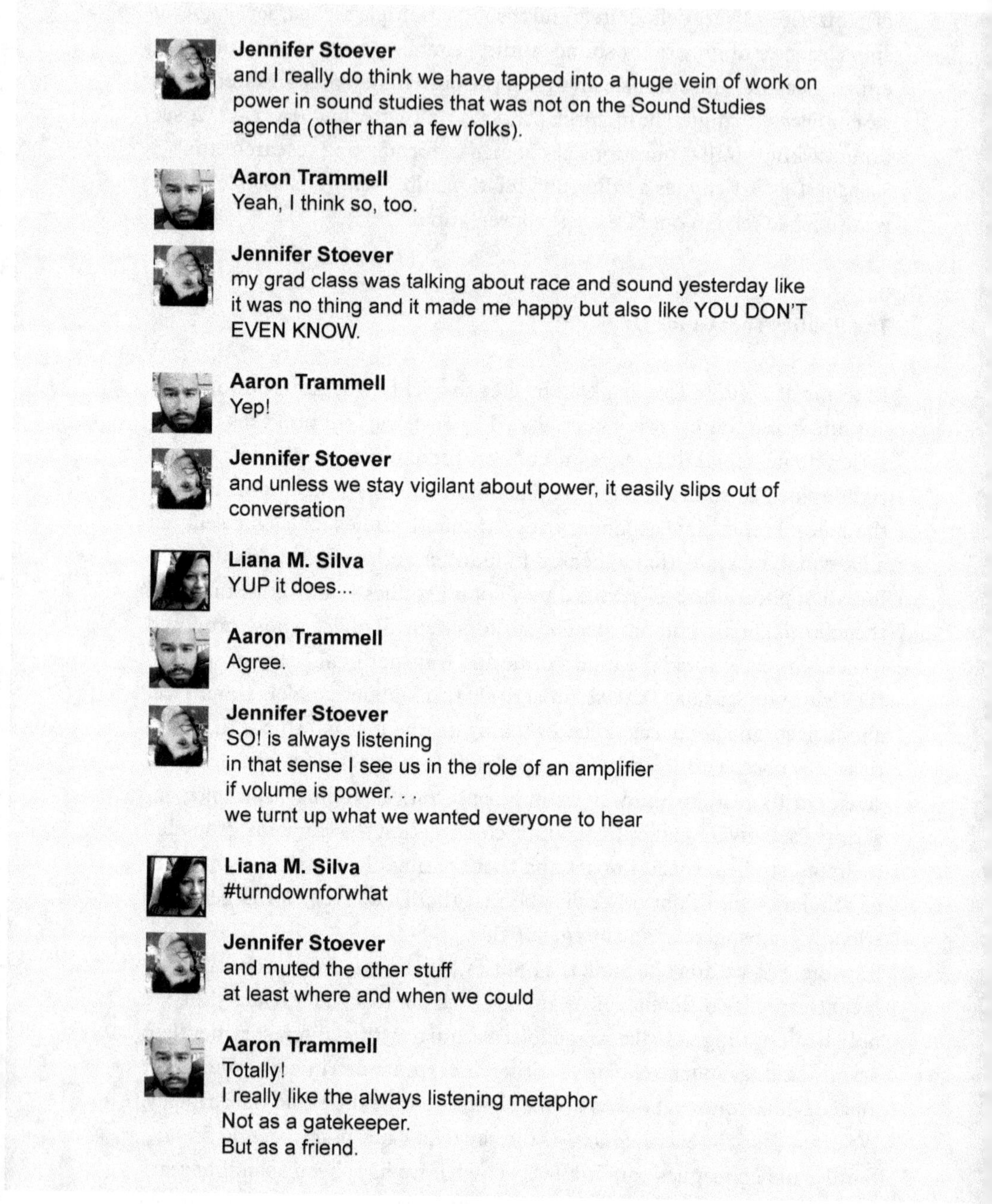

Jennifer Stoever
and I really do think we have tapped into a huge vein of work on power in sound studies that was not on the Sound Studies agenda (other than a few folks).

Aaron Trammell
Yeah, I think so, too.

Jennifer Stoever
my grad class was talking about race and sound yesterday like it was no thing and it made me happy but also like YOU DON'T EVEN KNOW.

Aaron Trammell
Yep!

Jennifer Stoever
and unless we stay vigilant about power, it easily slips out of conversation

Liana M. Silva
YUP it does...

Aaron Trammell
Agree.

Jennifer Stoever
SO! is always listening
in that sense I see us in the role of an amplifier
if volume is power. . .
we turnt up what we wanted everyone to hear

Liana M. Silva
#turndownforwhat

Jennifer Stoever
and muted the other stuff
at least where and when we could

Aaron Trammell
Totally!
I really like the always listening metaphor
Not as a gatekeeper.
But as a friend.

FIGURE 4.5 Discussing the need to "stay vigilant about power" and race and how we see *Sounding Out!* as an amplifier, a listener, and "a friend."

When *Sounding Out!* plots our publishing calendar, we think about academics and nonacademics. About senior and junior scholars. About graduate students. About women. About people of color. About people at various points along the spectrums of sexuality and gender. About specialists and nonspecialists. About alt-acs and independent scholars. We actively seek artists, sound professionals, curators, musicians, DJs, game designers as practitioners, experts, and theorists. While we cannot promise perfection, we do promise perpetual vigilance; our open submission policy, comments section, and social media platforms enable our commitment and allow our readers to assist in this process. We host diverse conversations not as a vague gesture toward inclusion or a specious invitation for "others" to join a preexisting conversation, but rather as a blueprint to construct a lasting, interactive community that values a variety of epistemologies, welcomes diverse and multimodal forms of rhetorical address, and involves and connects people rather than compiles an abstract, empty referent. While the online format enables *Sounding Out!* an unprecedented reach and a much more democratized distribution network, our *Sounding Out!* community thrives through a digital rendering of an analog sense of affect, as our survey results reveal in the subsequent sections.

Blogging and/as Community and Platform

"Blog" is a key term for the editorial team. It is literally embedded in the URL of every webpage of the site, sure, but that embeddedness is emblematic of how "blog" is more than just a noun for us. Blog is ethos, rhetoric, and form.

For us the term "blog" best captures the productive tension *Sounding Out!* creates between "journal" and "magazine," "seriousness" and "play," "academic" and "public," with the added layer of sound and visual media capabilities a digital platform enables. Our commitment to the term is both practical—"soundstudies.com" was already taken, so "soundstudiesblog.com" seemed like the next best address—and tactical, freeing us to experiment in ways that might "tarnish" a journal's reputation or frustrate a magazine's readership. Furthermore, the close association of "blog" with Internet 2.0 immediately signaled different expectations to our writers and readers—namely that there will *actually* be sound embedded in the writing in a meaningful way.[16] For many of our writers, just knowing *Sounding Out!* offers them the capability to embed sound significantly shifted how they approached their work.[17] Although many of our posts appear at first glance

Liana M. Silva
and i'd be interested in exploring more the concept of the blog, to see how that fits into what we're saying/experiencing.

Jennifer Stoever
i think we cling to the word blog because to us it signifies a kind of freedom and flexibility to reinvent and evolve that the other terms don't seem to.
what the hell is an "online magazine" anyway?
you know?

Aaron Trammell
Yeah.

Liana M. Silva
it would be neat to situate SO as blog in a broader conversation about blogs.

Aaron Trammell
I'm more practical with it.
We're literally stuck with it.

Jennifer Stoever
true

Aaron Trammell
It's part of our identity.

Jennifer Stoever
it is in our name

Aaron Trammell
Like it or not
So lets not shun ourselves for it
Lets embrace it and love it.

FIGURE 4.6 Our discussion regarding concerns over the term "blog."

to be written posts that include sounds, our editorial experiences with writers and their responses to our survey (discussed below) reveal a much more complicated process at work. Using a multisensory digital genre enables folks who are writing for online platforms to "think with" sound and image in new ways, from the very inception of an idea, an advance that has significantly shifted the writing itself. Furthermore, the flexibility of the medium (e.g., add a widget on the sidebar, review the list of categories for the posts, embed an audio file in a post, start a real-time discussion in the comments) allows us to constantly reinvent how writing about sound studies looks and sounds. A tour of our readily accessible back catalog will show how much we've grown and how our editorial sensibilities have developed, particularly in using the visual as a sonic medium online. The categories in themselves allow us to index a field that is no longer burgeoning but still changing and responding to current events.

Over the years we as a team have debated whether to move away from a blog format, especially as we considered how changing our nomenclature to "journal" would give us a certain legitimacy with academic audiences outside of our readers and writers. Shifting the title to "journal," however, shortchanges the many others who are doing great—intimate and immediate—work with blogs. We lose in spirit when we identify as something we are not. So we revisited our charge and decided that we *are* a blog. We didn't need to be a journal: there are now journals publishing work in sound studies, and we recognize that some scholarship benefits from the slow approach of a print journal. We do not see blog in opposition to journal; all three of us regularly read, publish, and cite print scholarship. *Sounding Out!*, however, provides a new space for a different kind of scholarship, because it

- is improvisational,
- responds to current events, and
- mediates between academic scholarship and nonacademic responses and the praxes of both.

More importantly, *Sounding Out!* is not just a different format for academic scholarship; it forces readers and writers to consider the way the work is produced. As blog editors, we work closely with writers about their writing, we communicate constantly with them regarding revisions, we promote tirelessly their work via our social media profiles, and we ultimately see the creation of the multimedia blog post as a collaborative effort. We do not leave our writers alone. We are there via email, tweet, or even in conversations happening in the comments to a draft. The blog, in essence, is not just a space online to post work; it also becomes a work ethic where we develop and produce each other's work. We write, we comment, we post, we listen.

It is also important to point out that hosting a blog requires a kind of work that journals often take for granted: we must vigilantly tend to our presence in the World Wide Web. Every post is carefully tagged not just for the sake of our readers but also for the sake of connecting *Sounding Out!* to online searches around the world (and permanently archiving *SO!*'s participants). If the categories are an interior indexing mechanism (like a table of contents), tags are echoes bouncing back into the internet. (They literally help to index us for search engines.) This careful attention to categorization also helps us stand out in search results. The essential work of search-engine optimization—categorization, headline building, index management, and layout (the mundane tasks of web maintenance)—is seldom recognized as valuable labor by the academy. We work hard to make sure that *Sounding Out!* blog entries appear as relevant search results for anyone looking for insightful reading on sound. These tasks exemplify the best practices in digital publishing and make clear some of the many ways that digital publications can be evaluated.

And this is where form begins to trump content when it comes to the label "blog." While much academic energy expends itself in debating whether a blog "counts" as much as print scholarship, scholars and administrators alike pay very little attention to the structure and function of a blog as digital craft of a radically new order.[18] Precisely because of its radical affordances, the debate over "public scholarship" somewhat belittles the participation enabled by the blog form: How else could two graduate students of color and a first-generation, working-class junior academic (two women, one man) establish a publication that has made such deep imprints in the field? The blog threatens established hierarchies and allows for new voices to slip in and expand discourses that previously have been hermetically sealed. The blog can do this because it relies on affective affinities between its editors, writers, and readers, as opposed to the economic and patriarchal affinities of the print journal, the established hierarchies of rank and review in the academy.

We, as editors of *Sounding Out!*, consciously choose "cred" over "credit," particularly when working with our authors. The long hours spent editing (and laying out) each post are uncredited, and many colleagues assume the vetting to be less rigorous than the work of peer review for a journal. No course releases are provided for our work; no grants have ever been

Jennifer Stoever
I still like the idea of a platform
in the old sense of the word, not just digital

Liana M. Silva
that's precisely it: we provide a platform.

Jennifer Stoever
we built our own platform from which to speak, which is difficult to do.
but we also recognized that listening is just as important as speaking--and we cultivated a community of listeners

Aaron Trammell
I love that!

FIGURE 4.7 Our discussion of what the term "platform" means to us as an editorial team.

awarded to us; and *Sounding Out!* was given only a couple of sentences, under "service," in Jennifer's official tenure case. This is not to say, however, that we haven't accrued other important benefits from our labor, such as much higher visibility, more invitations to editorial boards and collectives, and wider national and international networks than are available to many early-career scholars and alt-academics. Most meaningful, however, to the three of us is the strong sense of "cred" we have steadily earned within our community by maintaining pleasurable professionalism and a superlative internal standard. Those who become part of our community come to rely on us, and in turn they do what they can to spread the word.

While the "always-on" feel, conversational tone, and time-sensitive publication of *Sounding Out!* certainly have helped build this actualized community, we as editors have built it link by link by link.[19] Linking is not terribly sexy labor—both web users and university administrators take it for granted—but to us it feels like breathing, an almost unconscious practice necessary to animate the entire structure. For example, our decision to embed links rather than use footnotes was tactical rather than stylistic (even if it runs counter to the style guides we memorized as undergraduates), enabling us to further embed ourselves within conversations about sound occurring on the web. Links perform the function of citations, but they also shape search-engine results; according to Tim O'Reilly, the more "prolific and timely" a blog's links (and "self-referential" within a community), the more the process of "bloggers paying attention to other bloggers magnifies their visibility and power."[20] And sure enough, after seven years of tireless linking between blogs, journals, universities, and social media sites (over 17,400 tweets as of July 2016!), if you enter the search term "sound studies" in Google (as we asked our survey respondents to do), *Sounding Out!* comes up in the first five entries, often in the top three, just under the Wikipedia entry (which lists us) and Sterne's canon-making *The Sound Studies Reader* as key resources for the field. Importantly, our location means that just about *anyone* looking up sound studies—from undergrad to sound professional to grad student to colleague to grandparent—will come across *SO!* and its interventions regarding sound, social difference, and power early on, insuring such inquiry will become—and remain—the heart of the field. Our hard-fought Google rankings represent something far more important than winning results of a popularity contest or nice evidence of "reach" for university administrators perusing our tenure files; it reveals the literal and figurative "platform" we have worked to build for ourselves and our community. Again, we've developed "cred" in lieu of "credit."

When we met in a humid apartment in upstate New York to plot a sound studies blog back in 2009, one of our key goals was to provide indelible visibility to the top-notch contributions we knew were being made to sound studies by scholars of color, graduate students, junior scholars, and other disempowered groups in academia, so that their role in building this growing field could not be erased, ignored, silenced, hijacked, buried, or claimed by others better positioned by social and institutional privilege and its attendant cultural capital to gain conference spots and find publishers for their work. There is solidarity in the affects produced by giving voice, making visible, and—above all else—listening. As Sundén argues, "The ways in which we imagine and feel for technologies matter," so we decided to build our own site and to do so in a way that celebrates the people and the scholarship perpetually at the fringes of most fields, but especially those involving technology and music.[21] *Sounding Out!*'s consistent publication and voracious linking structure created the platform; we then combined well-written, cutting-edge, quality scholarship with participatory social media; targeted blogrolling; in-person conference marketing and social events; active recruitment and developmental editing; and colleagues' support through retweets, shares, pings, and traditional citation to create an ever-growing community of listeners surrounding it. The blog listens, it breathes, and it provides a center to anchor the precarious labor of fringe scholars who might otherwise be swept away in the market-driven and opportunistic frontier of the digital rhizome. Industry practitioners, graduate students, and independent scholars have the most to lose by blogging, but they also have the most to gain when it is done right. We strive to support these vulnerable scholars in any way we can.

Survey

Because our goal has always been to foster a greater degree of affinity around the topic of sound studies, we felt an essay of this kind would be incomplete without affording some insight into how *SO!*'s primary participants understand this sense of community (rather than just speculating or assuming our theories always rang true). We wanted to listen to the participants in our community so that we best represent ourselves as the collective, posse, and crew we are. Our blog would not be as successful—or as fun—as it is without the labor of the writers who contribute week after week. To better understand how *Sounding Out!* serves its contributing network of digital scholars and activists, we conducted a survey that queried for qualitative data regarding the publication's reputation, circulation, reception, and editorial process. We chose not to administer our survey anonymously due to the level of detail we requested—essentially, we would have been able to identify respondents anyway—and we sent it to every guest writer who has written for *Sounding Out!* since the site's establishment in 2009. In total, we received twenty-four responses from a total pool of one hundred participants. We administered three follow-up questions to these twenty-four respondents in late January 2015 and received twelve responses.

We coded the results using a grounded theory methodology that allowed our data to speak for itself and reveal a set of relevant categories.[22] During the coding process, we compared results and selected emerging themes and categories as well as identifying several interesting (yet understated) categories to unpack in this essay. Our aim here is to highlight a sense of consensus about *Sounding Out!* as well as to provide some insight into how this consensus has been challenged, for instance in terms of the editorial process or our place in the digital humanities universe.

We also chose not to make our survey anonymous because we felt that personality and profession would play heavily into the ways in which our respondents would consider the rife political nature of these questions. As such, we wanted to be able to weigh and acknowledge how responses were relative to a particular professional positionality. We also wanted to better understand and credit the labor of our contributors. We posed the following eight questions (or prompts) to our survey respondents:

1. How would you describe *Sounding Out!*? How do you see it in relationship to the digital humanities community?
2. Describe your personal involvement in *Sounding Out!*
3. How was your experience of the editorial process?
4. Please describe your experiences with any or all of our social media platforms (Facebook, Twitter, Tumblr).
5. Has *Sounding Out!* aided and abetted your scholarship, art, sound work, and/or any other capacity? If so, please tell us how.
6. What has been the best part of being involved in *Sounding Out!* over the years?
7. What do you think that *Sounding Out!* could do better?
8. Any final thoughts you'd like to share with us?

We then asked all respondents these two follow-up questions:

1. When you search for "sound studies" in Google, where is *Sounding Out!* in your results?
2. Very basically and honestly, why did you publish your work on *Sounding Out!*?

The respondents had the opportunity to respond online between April and May 2014, just in time for the blog's fifth anniversary, and to follow up in January 2015.

In terms of broad trends, respondents commented about our editorial acumen, pointing out how rigorous the editing process is and how rewarding it is at the end. Our respondents saw *Sounding Out!* as a resource, a hub, and a platform, but very few saw it as a "blog," judging by their avoidance of the word itself. Many also follow *SO!*'s Twitter feed, which they enjoyed both for its informative and for its personable qualities. Respondents used words that suggested they *feel* an affective connection to *Sounding Out!* and the community it fosters: we noted the recurrence of words like "helpful," "connect," "accessible," and "isolation." Survey respondents also noted that they came to the blog to keep up with the field and that, in various ways, it enabled them to feel part of a wider community. In the following subsections we discuss the results in detail, focusing on how respondents felt a connection to sound studies, understood our editorial process as peer making (not just peer reviewing), defined themselves as writers AND readers, and actively engaged with *SO!*'s microinterventions on Twitter.

Connections with the Discipline

Although we think of the website, our bloggers, and our readers as the *SO!* community, we also see ourselves as part of bigger disciplinary communities, part of sound studies, and part of digital humanities. Because we do not affiliate with an institutional structure to house our work—and have received no external funding—we rely on connecting with other scholars to feel like part of an academic network. Our bloggers agree that they feel connections with those disciplines and with each other through *Sounding Out!*

In the survey, several respondents across ranks mention how they see *Sounding Out!* as a way to stay involved with sound studies. For example, Meghan Drury (a graduate student when she took the survey) mentions that "*Sounding Out!* provides an important digital resource for sound scholars in the U.S. and worldwide ... the posts on *Sounding Out!* stimulate my intellectual development and encourage me to think about sound scholarship in new ways." For Drury, the blog provides not just reading material but also professional development within the field. Associate Professor Priscilla Peña Ovalle, who describes herself as a scholar in a field "adjacent" to sound studies, states that writing for and reading the blog become for her a way to stay in touch with the field. Kariann Goldschmitt, now an assistant professor, shares that "the network of thinkers involved in the site is really exciting. Whenever I run into people at conferences, we have a deeper understanding of each other's work. That's incredibly rewarding." Reading the blog becomes a way to perform scholarly community, to understand the work of other sound studies scholars by reading their work on the site and sometimes engaging them in conversation via social media or email. *Sounding Out!*, in this case, is a meeting ground for ideas and scholars. And bringing scholars together to talk about anything is like herding cats, #humblebrag.

Regarding digital humanities, some of our respondents were unsure about their understanding of the term "digital humanities"—or if *Sounding Out!* qualifies based on a rubric of "big data"—but others believed that the blog exemplifies what a project-based digital humanities community can be. For example, recent PhD and now assistant professor Steph Ceraso points out, "I think that the *Sounding Out!* community is a wonderful example of what the DH community strives to be: a welcoming space for new ideas and diverse voices, a community that encourages collaboration, an open community that freely creates, shares, and builds upon ideas, and a community that is always respectful and generous to its members." She stays away from references to the digital platform and instead focuses on the possibilities of a space that brings together a diverse group of scholars and practitioners, a situation particularly meaningful for her as, at one time, the only student in her department dissertating on sound studies. Meanwhile, professor and curator of the Rose Goldsen Archive of New Media Art Timothy Murray connects *Sounding Out!* with DH conversations about "hack vs yack": "*Sounding Out!* is a forceful, performative blog that links makers, thinkers, and listeners in the critical involvement of studying sound." Overall, the survey responses show that *Sounding Out!* offers an understanding of community aligned with the social connotations of the digital, but one whose meaningfulness and pleasure are enhanced through the relationships *Sounding Out!* enables and strengthens IRL (in real life).

Editorial Process

Commonsense undertones carried by the word "blog" can betray the editorial labor that goes into each post, which is connected to how editing is perceived in academia overall: as begrudging necessity rather than pleasurable community praxis. In other words, editing is considered service, an undervalued category of scholarly work. Ever-dizzying work schedules and publication expectations in the humanities and social sciences have made editing a far less collegial practice, one performed quickly, quietly, and with less-than-desired amounts of interchange. The traditional blind peer review model, particularly when combined with the work speed-up, can lead to a one-sided exchange of punitive comments rather than productive feedback; after all, the same busy colleagues with little time to form a writing group are the same folks tapped, often unexpectedly, to perform uncompensated ad hoc editing for professional journals. More often than not the cloak of anonymity, proposed as a meritocratic guarantee of objectivity and quality, masks curtness and flat-out rudeness as reviewers brusquely pass judgment rather than leaving comments intended to develop the piece. The current traditional editorial model leaves writers bereft of mentorship or critical dialogue about their work at perhaps its most crucial point; even if a writer discusses readers' feedback with their editors, it is mainly in terms of "what needs to be done to satisfy the readers" to get the piece published. There is rarely, if ever, another read beyond copyediting. Not merely a missed opportunity for productive exchange, traditional blind peer review (as it is currently practiced) actively fosters isolation.

Because we consider the community-building function of *SO!* as its primary purpose, we prefer the verb "host" to describe how we disseminate scholarship, rarely using the word "publish"—even if the button we click on WordPress says exactly that. Our respondents, too, emphasized the role of the blog as a host for sound studies scholarship. For example, Assistant Professor Tom McEnaney mentions that "*Sounding Out!* is the preeminent place to go—in print, or online—for innovative work in sound studies." His comment draws attention to the blog as a location where readers come to find new work in the discipline. Goldschmitt states, "*Sounding Out!* is an important forum for discussion and nascent scholarship." Professor Karl Swinehart adds, "*Sounding Out!* is an important venue where scholarly work within sound studies is presented in a multimodal format and in an idiom that is accessible across disciplines." The references to the blog as "venue," "forum," or other site to encounter work in sound studies draw attention to how the blog provides writers with a platform to share their work while connecting them to readers eager to hear what they are working on.

For *Sounding Out!* to host exciting writing and new research, as editors we work as cohosts throughout the editorial process. Combining Kathleen Fitzpatrick's "peer-to-peer review" model, in which editors and writers are known to each other, with the praxis of developmental editing more common to popular print media and trade presses, *Sounding Out!* pursues editing's community-building possibilities through practices that build trust and accountability through communication.[23] As McEnaney states, "As a writer, I found the editorial process intensely engaged, and incredibly helpful." In addition to providing writers with "extensive freedom in style and locution," as editors we operate as a medium connecting writers to others through tone and address. Murray recognizes that we work "assiduously with bloggers to keep posts accessible to the broad audience of the blog." Not only does our credibility as a resource lie in the editorial work we do, but we also believe that our "peer-to-peer" editorial relationship provides an important foundation to the blog, improving the tone and quality of the writing and benefiting the sound studies community itself.

Hosting, of course, does not mean the material is presented "as is." Quite the opposite. We work extensively with our guest writers to help them develop their ideas and address questions they may not have considered. Associate Professor Marci McMahon, for example, muses about the editorial process: "This was actually much tougher than writing a standard scholarly journal article. The editorial process is rigorous and the expectation to write a smart, pithy, and clearly understandable piece in 1,500 words is not easy to do. The editorial staff is tough and demands a lot from your work!" Contrary to academic journals, we do not expect "finished" essays the first time around, and we tell our writers their drafts will go through at least two rounds of edits, the first of which will be developmental.

In addition, we are open to unsolicited contributions and have a very low rejection rate, something we take much pride in, especially given how many respondents remarked on our quality and high standards. Sometimes our editorial collective will reach out to writers for posts, and other times writers will pitch an idea to us to see if we would be interested in the full draft. Once assigned to a project, a member of our editorial collective reaches out to the writer, making themself available for questions, pitches, and quick reads of difficult passages. Our guidelines explicitly ask for a first rather than a final draft, enabling writers to send early idea-driven versions that open up possibilities for dialogue between writers and editors in successive drafts. Rather than issuing global comments about a piece and then leaving the writer to decode them in solitude, *Sounding Out!* editors use Word or Google Docs to leave in-text notes that writers respond to directly, another form of community by microintervention: we are asking questions, recommending sources, leaving observations inspired by the draft, suggesting other scholars to contact, sending relevant links, explaining why we made a particular change, making connections to their own work, commending a particular point or turn of phrase and pushing for more. Dropping in jokes, emoticons, and emojis along the way, we're finding unique ways, in context, to imagine and discuss the next iteration of the post. Using the "track changes" function, editors also make grammatical, syntactical, and organizational changes directly to the text, carefully sculpting the piece's rhetorical flow and helping writers make new connections. Writers often work with multiple editors—one or more for each draft, all working on the piece toward the goal of publication—widening the margin-note conversation beyond narrow notions of expertise and ensuring each post will speak to multiple audiences. Jennifer often pairs graduate students and early-career scholars with editors in their field whom they have not yet met, so that they leave the editorial process with a new connection and a short-term working relationship that may lead to future information sharing and collaboration. The pleasure of meeting new people and strengthening network bonds is a key part of our model. Where magazines and other for-profit journals offer money, we offer community and connections—and therefore rigor and accountability. Writers and editors are thus accountable to each other and each has a stake in a piece's successful publication.

Our survey respondents agreed that the editorial process is long, and those who have published in traditional academic journals often compare the process with peer review—the result being that Sounding Out! always emerges as more detail-oriented and exacting. However, they don't see this as a negative thing. One of our writers, a PhD candidate, described the process as "a little too hands-on," but most of the other responses saw the process as essential to their pieces. A graduate student when he worked with us (now a PhD and writing center director), Airek Beauchamp states, "The editorial process was rigorous and ultimately transformative, in the best possible way." Peña Ovalle mentions that "the editorial process is exemplary. Thanks to the incredible feedback, my work was pushed and polished in a way that exceeds the standards of many traditional scholarly print publications." This is not to say that editors at academic journals are not careful or detailed; however, we acknowledge that developmental editing is time consuming and "inefficient"; most scholarly journals cannot find enough willing editors of this stripe with field expertise, particularly with dwindling budgets. And, certainly, both editors and writers must constantly balance Sounding Out!'s pleasures with the knowledge that our unpaid work may likely go unsung and uncredited by our institutions and supervisors.

However, our guest bloggers find our process pleasant and helpful, and they notice that we do, too.[24] While our labor remains "free," it is also freely given—and we strive to ensure the relationships we build give back. In contrast to how some authors may describe working with an editor as grueling, our writers for the most part enjoy working side by side with their editors. For example, PhD candidate Enongo Lumumba-Kasongo states, "I have thoroughly enjoyed the editorial process. Aaron [Trammell] has been nothing but professional, timely, forgiving, and very thoughtful in his critiques and suggestions." This emotional connection helps establish Sounding Out! as a community, bringing writers back to write for us again and again. Ceraso articulates the connection: "From the start, I felt that [Jennifer Stoever] genuinely cared about each contribution." Our process has been especially helpful in increasing international communication in the field. Finnish PhD student Kal Ahlsved responds, "Since English is not my first language I am very thankful for the editorial patience. I really learned a lot about how to hold a thought and to follow a stream of thought." In addition to enjoying the editorial team's field knowledge and writing skill, writers notice—and respond positively to—the "patience," "enthusiasm," and, as

Lumumba-Kasongo puts it, "positive feedback and words of encouragement, something that actually makes a huge difference when being asked to rework something multiple times." To our surprise and delight, several respondents reported being inspired by our editorial praxis in their work in other venues, both on- and off-line.

Our editorial process brings out the "digit" in "digital," as Bratich would say, humanizing our community and making it feel realized rather than "imagined." The guest writers who responded to our survey do not see *Sounding Out!* as a gate that keeps them out of sound studies, but as representative of a group of people who are interested in developing their ideas, helping the quality of their writing and recording (our podcasters also go through this editorial process), and amplifying their work throughout our networks. We work hard to ensure that our writers—particularly junior faculty, graduate students, community workers, and artists—have a chance to share their ideas with a broader scholarly community, exciting new ideas that otherwise might have been rejected from traditional academic journals and set aside, perhaps forever. Because scholars burn out when they go unheard, we perform the emotional care-work of supporting our colleagues who stand at the margins of academia.

PRESENCE/PRESENT/IMMEDIATE: Social Media and Microinterventions

Building from our personal editorial relationship with writers, our social media presence has been integral to creating the kind of "big tent" sound studies readership we imagined for *Sounding Out!*, while potentially reaching people outside of academia like those in the art world, the sonic professions, and the friends and family of *Sounding Out!*'s blogging crew. As we discovered early on, merely placing information on the web does not build community in and of itself. To encourage a cross-platform community centered around but ultimately reaching beyond the "mothership," we worked hard to craft a distinct purpose for each social media outlet, a move that also enabled all members of the editorial collective to curate their own unique, but connected conversations. In other words, our Twitter, Facebook, and Tumblr feeds are not adjacent to the blog; they *are so!*

While our social media presence seems *Sounding Out!*'s most readily apparent community-building enterprise, the path toward a functional, connected network has been anything but clear. Each medium—WordPress, Twitter, Facebook, iTunes, Tumblr, Google Plus, Reddit, and a monthly

FIGURE 4.8 A greyscale version of the SO! logo (created by artist Dan Torres), which is emblazoned across all our platforms.

emailer that goes out to more than 1,100 subscribers—has its own conventions, protocols, and even audiences, and it took much brainstorming, trial, and error, to discover how to reach out effectively. Our respondents underscored this point as they shared how they connect with us on many social media platforms. As we played around with different social media profiles, *Sounding Out!* held fast to two main concerns: legibility and accessibility. We wanted to ensure that interested parties at each access point in *Sounding Out!*'s constellation of social media would immediately recognize our "digits" at work yet would also find unique information and conversations there. Such diffusion, we felt, would enable more mobility for the sound studies community—not being housed or affiliated with any one virtual location—and offer an increasingly diverse range of ways for interested parties to feel connected, share information, join conversations, reach out to each other, and spread the word through shares, likes, retweets, reblogs, "+1s," and up-arrows.

According to our respondents, Twitter is the platform where we make the most sound waves outside of WordPress. Inspired by Liana's microblogging as @Literarychica, Jennifer took on Twitter about a year after *Sounding Out!*'s founding, and she has steadily cultivated a feed of artists, scholars, presses, archives, organizations, programs, digital humanists, and public figures, a diverse well from which to retweet calls for papers and sound-related news, articles, events, releases, job listings, and media clips to *Sounding Out!*'s 5,415 Twitter followers (as of January 2018). Twitter folks can also subscribe to the curated list of more than five hundred "soundtweeps" to tap immediately into a more concentrated conversation regarding audio culture. Jennifer also regularly livetweets conferences, talks, speeches, art openings, and other cultural events of interest to *Sounding Out!* followers and passes on information gleaned in her own research in the field. Followers use the @soundingoutblog handle to ask Jennifer

FIGURE 4.9 *SO!* contributor Steph Ceraso spotted repping the blog by fellow *SO!* writer and recent PhD Tara Betts at the Feminism and Rhetoric conference in January 2014.

questions, crowdsource problems, pitch a post idea, seek knowledgeable parties, and share their own news and interesting web clippings for *Sounding Out!* to retweet.

These exchanges make a difference. Lumumba-Kasongo, for example, describes "a number of positive exchanges with other individuals who have learned about my research interests through tweets that were sent from *Sounding Out!*," including a moment when we mentioned his "piece on audio games to someone on Twitter who mentioned an interest in sound and games, and we ended up having a nice dialogue about some of my discussion points." The flow of conversation moves outward and in unpredictable ways. Jennifer frequently interacts with followers by asking questions, seeking writers, commending observations, asking for collaborations, engaging with memes and hashtags, cracking jokes, and calling out misinformation and/or bad practice. Finally, she regularly updates followers on *Sounding Out!*'s writer-related news like graduations, publications, promotions, performances, and travels, personalizing the community and building affinities within and without the always-expanding Team *Sounding Out!* There is some content crossover for the 3,935 (as of January 2018) folks who have liked our Facebook page, but with an increased emphasis on providing an archive of sound studies CFPs through Facebook's "notes" feature; images and lengthier informal updates from relevant conferences *Sounding Out!*

editors attend; and a community-building photo series that encourages readers to send in images of *SO!* stickers—paid for by us and distributed for free—that they spot around the globe.

Writers, Readers, Sharers

Our work at *Sounding Out!* is not limited to hosting content and tweeting news; we are always thinking of our readers as well. Indeed, many of our guest writers are regular readers of *Sounding Out!* and feel a long-term stake in the blog even after the editorial process ends. Many of our writers confessed that they continued to read the blog on a regular basis after their work was featured. Wanda Alarcón, a PhD candidate when she took our survey, describes herself as "reader, guest contributor, fan." The use of the word "fan" in this instance points to admiration of the blog and pleasure in reading it on a regular basis. Ahlsved mentions that he reads the blog regularly and that he often shares relevant pieces with his peers. Sterne says he "looks forward to reading it every Monday morning," referring to our first post of the week. Assistant Professor Jentery Sayers also admits to being "an avid reader." The regular but measured pace of the blog helps readers keep up with the content, with one or two new pieces a week and a podcast per month. However, the content reigns supreme; because the writers know how much care goes into each post, they are assured that every post is a well-written addition to the field.

The fact that writers continue reading, sharing, and interacting with the blog—be it through likes, comments, contributions to our annual Blog-o-Versary mixtape, or sporting a sticker or button—shows that they feel invested in the community of the blog. Readership is not a passive exercise but in fact supports the scholarship of other scholars. When asked to describe the best part of being involved in *Sounding Out!* over the past five years, Soundbox cofounder and Duke PhD candidate Mary Caton Lingold says that it has been "getting to know scholars from other institutions and being able to share work and ideas with them." Drury reiterates this feeling: "I have found it useful to learn about the work others are doing in the field." Bill Kirkpatrick, associate professor, sums up these ideas nicely, admitting, "The best part has been feeling like part of a community of scholars. I appreciated being invited to participate, and I like reading what others have to say." The responses indicate that reading is a way of enacting scholarly citizenship as well as keeping up with what's going on in *Sounding Out!*

Although pressures from the job market and tight tenure clocks demand

FIGURE 4.10 Every July we commemorate our first post by producing an annual collective mixtape with song suggestions from the year's contributors. Logo for our tenth-anniversary mix by Jennifer Stoever and Aaron Trammell.

an ethic of writing from us as scholars, it is important to remember that reading is an integral part of the community loop. Good scholarship means writing *and* reading, and even sometimes writing an addendum in the comments about the post. The ethic of readership and participation fostered by *Sounding Out!* is, in fact, a solution to the manifold academic predicaments that have become readily apparent in the past thirty years. If we are to survive *as a profession*, we must rise to meet the demands and opportunities of today's new media platforms. As Clay Shirky articulates, "Media is a triathlon; it's three different events. People like to consume, but they like to produce, and they like to share."[25] We must become participants who read, write, and offer timely feedback to others in the field on a regular basis.

Conclusion

So in the end, you probably don't need to read between these lines to know we also do it for love.

And, in a sense, love is the affect that has sustained *Sounding Out!* and its affiliated network for the past five years. As social theorist Michael Hardt suggests, although the production of value from affect is often exploited by patriarchal and capitalist institutional forms, there exists a tremendous potential for affective labor to subvert dominant institutional configurations.[26] To this point, our firsthand experience and survey data show a thriving digital community that is paradoxically treated with apathy by the

Jennifer Stoever
i think we also don't find what we do taxing because the three of us have ALWAYS KNOWN we were gonna have to hustle. it was zero surprise.

Liana M. Silva
agreed. because #noprivilegehere

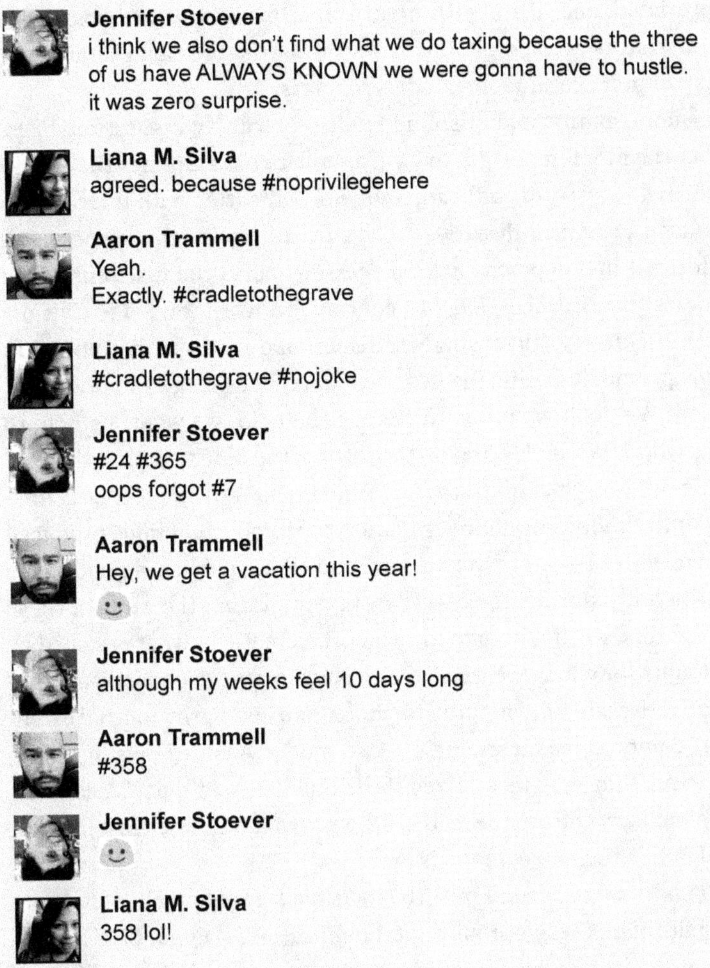

Aaron Trammell
Yeah.
Exactly. #cradletothegrave

Liana M. Silva
#cradletothegrave #nojoke

Jennifer Stoever
#24 #365
oops forgot #7

Aaron Trammell
Hey, we get a vacation this year!

Jennifer Stoever
although my weeks feel 10 days long

Aaron Trammell
#358

Jennifer Stoever

Liana M. Silva
358 lol!

FIGURE 4.11 Discussing our feelings about the tensions of love and labor as first-generation college students and "nontraditional" scholars in academia, marginalized by various intersections of race, gender, and class.

bureaucrats and administrators with whom we work. What goes unsaid, underappreciated, and seemingly unrecognized by these same bureaucrats and administrators is the digital network infrastructure that sustains our community of practice as sound studies scholars.

Furthermore, as the field of sound studies inevitably institutionalizes, it will be all the more important to have a vehicle that amplifies the granularity of the field and wards off status-quo normalization with increasingly radical linkages, particularly between the humanities and the sciences. But whether located in a department or dispersed across the disciplines, the sound studies that *Sounding Out!* will continue to work toward is civically engaged, participatory, increasingly transnational in scope, decolonial in theory and epistemology, and invested in applied knowledge and praxis-as-intervention. We don't want just to change the field, we want the field to change the world. We are betting on the form of the blog to do just that.

Although we find this infrastructure fundamental to our scholarly mission and our livelihood as public academics, the intellectual value produced from our collective labor is diverted into traditional publishing endeavors such as print journals and books. Far from denigrating the value of these traditional forms, we aim here to locate a problematic in what is valued by the institutions for which we work and to suggest that the mostly uncompensated affective labor of blogging is "more than just a print journal extension" or a "compromise technology"—two modes Ashley Dawson rightly critiques—and it must be recognized if the imbalance of today's academic publishing industry will ever be rectified.[27] So even though we did it for love, our digital publishing honeymoon is over.

We will continue to "sound out" the invisible lines of practice that constitute our site and other rigorous digital publications. Digital platforms—conjured into existence by a need for connection and the immediacy of scholarship on topics at hand—must be seen for what they are: the new configuration of the academy. And, as such, the work of editing (developing scholarship *and* community) must come to be valued by our institutions as much as the act of writing. There must be a recognition that reliability and trust stem from rigorous editorial processes as much as they do prestigious titles. And, perhaps most fundamentally, the microinterventions (tweeting, retweeting, linking, soliciting, challenging, and connecting) necessary to running a successful publication must be recognized as valuable labor in this new network and compensated with pay, positions, and prestige.

Sounding Out! continues to reward both us and the community, and this sustained sense of pleasurable community contact keeps us engaged on a fundamental level. We believe in the community effort that has both constructed and supported us, and we are proud to have seen terms such as "reader," "fan," and "inspiration" repeated in the survey results. We're in this together, and we must start the process of recognition by collectively—and loudly—revealing to our friends, colleagues, bosses, advisors, deans, provosts, and interested peers the affective labor practices that constitute our network, so that they can build awareness in turn about how much damn work goes into digital publishing.

And we must start by making more mixtapes. Always more mixtapes.

NOTES

1 These three categories represent the varied positions of the editorial collective at the time of writing. Over the course of the blog's existence, Jennifer has become tenured, while both Liana and Aaron have finished their dissertations. Aaron has successfully completed a postdoctorate and obtained a tenure-track job, and Liana has served as editor of *Women in Higher Education*, as well as being a freelance writer and editor. She is now a secondary school educator.

2 This is, in fact, what happened in Jennifer's otherwise successful tenure case. Although she provided extensive materials documenting the formation, growth, and impact of *Sounding Out!* (with extensive digital examples), and her supportive department took the proactive step of procuring an outside evaluator strictly for her digital scholarship—whose letter commented very rigorously and favorably on *SO!*—the evaluating dean undermined these efforts and her digital labor by describing "the blogspot *Sounding Out!*" as "a valuable service to our academic community" and therefore only an indirect contribution toward her "multifaceted" case for tenure. These were all concerns raised at the 2013 Modern Language Association workshop on digital scholarship and tenure, where Jennifer and *Sounding Out!* were selected as case studies to help scholars and administrators think through blogging and tenure. A panel at the 2014 annual meeting of the Organization of American Historians focused on precisely this question, featuring five different historical bloggers who addressed whether they considered their blogs scholarship. Points of view were mixed.

3 cummings, "since feeling is first," *Selected Poems*, 99.

4 We use "counternarrative" here to signal our intellectual solidarity with critical race studies methodology, in which researchers use storytelling methods to legitimate the extensive experiential knowledge of marginalized peoples and center conversations about race and power sublimated by dominant nar-

ratives. Also, as Daniel G. Solórzano and Tara J. Yosso argue, the term "offer[s] a liberatory or transformative solution to racial, gender, and class subordination" (24).

5 As of July 12, 2016, *Sounding Out!* was one of seventy-seven publications that were available only online and had no pagination.
6 Drobnik, *Aural Cultures*, 10. See also Aparicio, *Listening to Salsa*; Johnson, *Listening in Paris*; and Kahn, *Noise, Water, Meat*, on avant garde art and radio. See Kun on American literary and musical audiotopias (*Audiotopia*); Moten on the black radical tradition (*In the Break*); Picker on nineteenth-century sound (*Victorian Soundscapes*); Trevor Pinch and Frank Trocco on the synthesizer (*Analog Days*); Rath on early American soundscapes (*How Early America Sounded*); Bruce Smith on Shakespeare (*Acoustic World of Early Modern England*); Sterne on nineteenth-century audio technologies (*Audible Past*); Thompson on modernity and architecture (*Soundscape of Modernity*). A handful of formative anthologies were released in 2004: Bull and Back, *Auditory Culture Reader*; Erlmann, *Hearing Cultures*; Mark Smith, *Hearing History*; Drobnik, *Aural Cultures*; and Cox and Warner, *Audio Culture*.
7 For more on the methodology of a field in transition, see Hilmes's "Is There a Field Called Sound Culture Studies?"
8 Williams, "Community," 75.
9 Bratich, "The Digital Touch," 307.
10 Sundén, "Technologies of Feeling," 147.
11 Sample, "What's Wrong with Writing Essays?"
12 Boon, "One Nation."
13 In particular, the Sound Studies Caucus in the American Studies Association, the Sound Studies and Radio Studies Special Interest Groups in the Society of Cinema and Media Studies, the Music and Sound Interest Group in the American Anthropology Association, and the Sound Studies Interest Group in the Society of Ethnomusicology have been key foundational professional groups. In 2012 and 2014 *Sounding Out!* cohosted "meet and greets" with the Sound Studies Caucus at the American Studies Association annual conference.
14 There is now a dedicated print journal, *Sound Studies*, whose first issue was published in 2015. Jennifer is on the founding editorial board, no doubt due at least in part to her work on *Sounding Out!*
15 This is a problem in the digital humanities in general, as McPherson addresses in "Why Are the Digital Humanities So White?"
16 Even with recent compromise measures such as the inclusion of a CD at the end of a text or sound clips on an online "tie-in" site, written pieces have largely had to stand alone, without a sonic dimension, however necessary it might be to the analysis performed. Referring to these sounds is like referring to a text absent from the bibliography.
17 "I could actually have audible examples to accompany my analysis" says one respondent to our survey. "When writing my guest posts I could think/write

along with audio/video samples in mind simply because I knew that it was possible and also because it was the expectation," says another.
18 See Cohen's "The Blessay" for a distillation of this debate, particularly concerning writing at the intersection of journalism and scholarship.
19 We take the term "always on" from boyd's "Participating in the Always-On Lifestyle," in which she discusses the pleasures of staying connected and suggests hacks to make an "always-on" existence less taxing. As she argues, "There's nothing like being connected and balanced to make me feel alive and in love with the world at large" (74). We agree.
20 O'Reilly, "What Is Web 2.0?," 41.
21 Sundén, "Technologies of Feeling."
22 See Charmaz, Constructing Grounded Theory, 3–4.
23 Fitzpatrick, Planned Obsolescence, 43. See our full editorial statement online at http://soundstudiesblog.com/editorial-statemen, and our mission statement at http://soundstudiesblog.com/sound-studies-blog/mission.
24 Taken from our survey.
25 Shirky, "Gin, Television," 239.
26 Hardt, "Affective Labor," 100.
27 Dawson, "D.I.Y. Academy?," 261, 271.

WORKS CITED

Aparicio, Frances R. *Listening to Salsa: Gender, Latin Popular Music, and Puerto Rican Cultures.* Middletown, CT: Wesleyan University Press, 1998.

Bailey, Moya Z. "All the Digital Humanists Are White, All the Nerds Are Men, but Some of Us Are Brave." *Journal of Digital Humanities* 1, no. 1 (Winter 2011): n.p. http://journalofdigitalhumanities.org/1-1/all-the-digital-humanists-are-white-all-the-nerds-are-men-but-some-of-us-are-brave-by-moya-z-bailey.

Bender, Daniel, Duane J. Corpis, and Daniel Walkowitz, eds. "Sound Politics: Critically Listening to the Past." Special issue, *Radical History Review* 121 (2015): 1–7.

Boon, Marcus. "One Nation under a Groove? Music, Sonic Borders and the Politics of Vibration." *Sounding Out!*, February 4, 2013. http://soundstudiesblog.com/2013/02/04/one-nation-under-a-groove-sonic-borders-and-the-politics-of-vibration.

boyd, danah. "Participating in the Always-On Lifestyle." In *The Social Media Reader*, edited by Michael Mandiberg, 71–76. New York: New York University Press, 2012.

Bratich, Jack Z. "The Digital Touch: Craft-Work as Immaterial Labour and

Ontological Accumulation." *Ephemera* 10, nos. 3–4 (2010): 303–18. www.ephemerajournal.org/sites/default/files/10-3bratich.pdf.

Bull, Les, and Michael Back, eds. *The Auditory Culture Reader*. London: Bloomsbury, 2004.

Chachra, Debbie. "Why I Am Not a Maker." *Atlantic*, January 15, 2015. www.theatlantic.com/technology/archive/2015/01/why-i-am-not-a-maker/384767.

Charmaz, Kathy. *Constructing Grounded Theory: A Practical Guide through Qualitative Analysis*. London: Sage, 2006.

Chow, Rey, and James Steintrager, eds. "The Sense of Sound." Special issue, *differences* 22, nos. 2–3 (2001).

Cohen, Dan. "The Blessay," May 24, 2012. www.dancohen.org/2012/05/24/the-blessay.

Cox, Christoph, and Daniel Warner, eds. *Audio Culture*. New York: Continuum, 2004.

cummings, e. e. "since feeling is first." In *Selected Poems*. New York: Liveright, 2007.

Dawson, Ashley. "D.I.Y. Academy? Cognitive Capitalism, Humanist Scholarship, and the Digital Transformation." In *The Social Media Reader*, edited by Michael Mandiberg, 257–74. New York: New York University Press, 2012.

Drobnik, Jim, ed. *Aural Cultures*. Toronto: YYZ Books, 2004.

Edson, Michael Peter. "Dark Matter: The Dark Matter of the Internet Is Open, Social, Peer-to-Peer, and Read/Write—And It's the Future of Museums." CODE | WORDS: Technology and Theory in the Museum, May 25, 2014. https://medium.com/code-words-technology-and-theory-in-the-museum/dark-matter-a6c7430d84d1.

Erlmann, Veit, ed. *Hearing Cultures: Essays on Sound, Listening, and Modernity*. London: Bloomsbury, 2004.

Ferrada, Sebastian. "SO! Reads: Norma Mendoza-Denton's Homegirls." *Sounding Out!*, October 26, 2015. https://soundstudiesblog.com/2015/10/26/so-reads-norma-mendoza-dentons-homegirls.

Fitzpatrick, Kathleen. *Planned Obsolescence: Publishing, Technology, and the Future of the Academy*. New York: New York University Press, 2011.

Hardt, Michael. "Affective Labor." *Boundary 2* 26, no. 2 (1999): 89–100.

Hilmes, Michele. "Is There a Field Called Sound Culture Studies? And Does It Matter?" *American Quarterly* 57, no. 1 (2005): 249–59.

Johnson, James H. *Listening in Paris: A Cultural History*. Berkeley: University of California Press, 1995.

Kahn, Douglas. *Noise, Water, Meat: A History of Sound in the Arts*. Cambridge, MA: MIT Press, 1999.

Kun, Josh. *Audiotopia: Music, Race, and America*. Berkeley: University of California Press, 2005.

Kun, Josh, and Kara Keeling, eds. "Sound Clash: Listening to American Studies." Special issue, *American Quarterly* 633 (2011): 445–59.

Laws, Catherine, ed. "On Listening." Special issue, *Performance Research* (2010): 1–3.
McPherson, Tara. "Why Are the Digital Humanities So White? Or Thinking the Histories of Race and Computation." In *Debates in the Digital Humanities*, edited by Matthew K. Gold, 139–60. Minneapolis: University of Minnesota Press, 2012.
Moten, Fred. *In the Break: The Aesthetics of the Black Radical Tradition*. Minneapolis: University of Minnesota Press, 2003.
O'Reilly, Tim. "What Is Web 2.0? Design Patterns and Business Models for the Next Generation of Software." In *The Social Media Reader*, edited by Michael Mandiberg, 32–52. New York: New York University Press, 2012.
Organization of American Historians. "Is Blogging Scholarship? 2014 OAH Annual Meeting." Youtube video. April 16, 2014. www.youtube.com/watch?v=T2kfm7eOdaw.
Picker, John M. *Victorian Soundscapes*. New York: Oxford University Press, 2003.
Pinch, Trevor, and Frank Trocco. *Analog Days: The Invention and Impact of the Moog Synthesizer*. Cambridge, MA: Harvard University Press, 2002.
Rath, Richard Cullen. *How Early America Sounded*. Ithaca, NY: Cornell University Press, 2003.
Sample, Mark. "What's Wrong with Writing Essays?" Sample Reality, March 29, 2009. www.samplereality.com/2009/03/12/whats-wrong-with-writing-essays.
Shirky, Clay. "Gin, Television, and Social Surplus." In *The Social Media Reader*, edited by Michael Mandiberg, 236–41. New York: New York University Press, 2012.
Smith, Bruce R. *The Acoustic World of Early Modern England: Attending to the O-Factor*. Chicago: University of Chicago Press, 1999.
Smith, Mark M. *Hearing History*. Athens: University of Georgia Press, 2004.
Solórzano, Daniel G., and Tara J. Yosso. "Critical Race Methodology: Counter-Storytelling as an Analytical Framework for Education Research." *Qualitative Inquiry* 8, no. 1 (2002): 23–44.
Stadler, Gus, ed. "The Politics of Recorded Sound." Special issue, *Social Text* 28, no. 1 (Spring 2010). https://read.dukeupress.edu/social-text/issue/28.
Sterne, Jonathan. *The Audible Past: Cultural Origins of Sound Reproduction*. Durham, NC: Duke University Press, 2003.
Sterne, Jonathan, ed. *The Sound Studies Reader*. London: Routledge, 2012.
Sundén, Jenny. "Technologies of Feeling: Affect between the Analog and the Digital." In *Networked Affect*, edited by Ken Hillis, Susanna Paasonen, and Michael Petit, 135–50. Cambridge, MA: MIT Press, 2015.
Thompson, Emily Ann. *The Soundscape of Modernity: Architectural Acoustics and the Culture of Listening in America, 1900–1933*. Cambridge, MA: MIT Press, 2002.
Williams, Raymond. "Community." In *Keywords: A Vocabulary of Culture and Society*. Rev. ed. New York: Oxford University Press, 1983.

05

BECOMING OUTKASTED

Archiving Contemporary Black Southernness in a Digital Age

REGINA N. BRADLEY

I didn't expect "OutKasted Conversations" to catch so many people's attention. It started out as a pet project, a way to celebrate the Atlanta, Georgia, duo OutKast's twentieth anniversary in hip-hop. OutKast, an acronym for Operating under the Krooked American System Too Long, heavily influenced my coming of age in southwest Georgia in the 1990s. Their music offered a blueprint for thinking about black southern folks' lives (and why they mattered) after the civil rights movement. OutKast introduced the world to the funkiness of what hip-hop could do in the South, opening doors for the complexity of southern black life—pain, pleasure, remembrance, and perseverance. OutKast's body of work gave young southern black folks the green light to embrace their experiences and carve out a space to recognize their own agency rather than dismiss it as a side effect of the civil rights movement. "OutKasted Conversations" moved that conversation and recognition past music into the digital realm, creating a digital site for teasing out how hip-hop can serve as a catalyst of change in the post–civil rights American South.[1]

The Project Premise

"OutKasted Conversations" started as a lively conversation with friends over lunch about hip-hop albums celebrating their twentieth anniversaries in 2014. Names of now-iconic albums were thrown across the table over our lunches: Notorious B.I.G.'s *Ready to Die*, Da Brat's *Funkdafied*, Warren G's *Regulate . . . G Funk Era*, Scarface's *The Diary*, and Nas's *Illmatic*. My colleagues/friends—Bettina Love, Emery Petchauer, and Christopher Emdin—were most vocal about their excitement around the festivities regarding *Illmatic*'s anniversary. The album's anniversary would be well acknowledged, including a documentary *Time Is Illmatic* and a live orchestra performance of the album at the Kennedy Center for the Arts in Washington, D.C. For my colleagues, mostly reared in the Northeast, *Illmatic* represented the angst of growing up black at the end of the twentieth century. *Illmatic* provided artistic context for the socioeconomic disparities and strife affecting black urban America in the 1980s and early 1990s in the aftermath of the Reagan administration. Granted, *Illmatic*'s sophistication lies in its cross-section of jazz aesthetics and gritty, street storytelling, sonically and culturally pulling from the trope of New York as a hard and bustling city. For example, the consistent use of jazz piano, turntable scratches, and the sound of rustling subway trains along their tracks make *Illmatic* a masterful demonstration of hip-hop's function as a site of urbanity and contemporary black culture.

However, I didn't share the same level of excitement as my friends because my love of hip-hop didn't come of age in the Northeast. When Nas asked "Whose World Is This?" or declared a "New York State of Mind" (two tracks from *Illmatic*), I was not his intended audience. I was a country girl from Albany, Georgia. Tractors commanded the roads dusted over by dirt coming from fields of cotton, corn, and melons. Red clay was never idle if white shoes were nearby. Noisy cicadas and crickets fussed at each other early in the morning and late at night. OutKast was prominent on my playlist, helping me work through and recognize what it meant to be young, southern, and black. Many young black southerners used OutKast to find a voice amongst the murmurs of the past and present that flowed in and out of our everyday lives. Southern hip-hop provided a space for recognizing the complexity of a more current moment of black southern identity: coming to terms with the strides and shortcomings of the civil rights movement while taking joy in being young, southern, and black. OutKast demonstrated young southern blacks could dance and critique, laugh and mourn, and carve out space for unorthodox perspectives. OutKast's body of work offered a type of sensibility that catered to my southernness. They offered a rich sonic tapestry of historical southern black sensibilities—ring shouts, blues, gospel choirs and the black church, for example—while using hip-hop to establish their knowledge of self and while complicating the context of the South in hip-hop culture. Although urban southernness is more embraced in hip-hop today, OutKast introduced the possibility of the South as a contemporary and urban space. They cataloged Atlanta using a hip-hop hybrid of lyricism, spoken word, gospel, and funk music. OutKast signifies on rural and urban southern tropes to acknowledge the possibility of young southern blacks being able to carve out space within hip-hop while sustaining a narrative parallel to (not submissive to) the civil rights movement. I was excited for OutKast's reunion tour in honor of the twentieth anniversary of *Southernplayalisticadillacmuzik*. I was never old enough to go to a live performance when they were actively touring. Seeing a reunion performance (or three) was at the top of my list.

My initial premise for "OutKasted Conversations" was to create a space to celebrate OutKast's overall dopeness. I wanted to recognize their music and artistry as innovative and critical to hip-hop's development as a culture. As a scholar and member of the post–civil rights black South, I set out to celebrate OutKast's accomplishments and center them in more critical conversations taking place in (new) southern studies and hip-hop studies. In addition to providing a critical backdrop for thinking through OutKast's reunion tour, I also wished to push discussions into the multiple facets that OutKast covers in their work, including race, gender, education, economics, spirituality vs. organized religion, sexuality, and identity in the post–civil rights South. And, like OutKast, I wanted to extend the conversations we had about their work outside of cafeteria tables and back into mainstream discussions of hip-hop.

YouTube provided a platform to update the cafeteria-table talk trope: its easy access invited viewers to "pull up a chair" to the conversation, (re)introduce themselves to OutKast's music, and discuss how they can be positioned in hip-hop and in the academy. Hosting the series on YouTube simultaneously archived the discussion and pushed back against the way one listens to and critically engages with hip-hop in digital spaces. It served as a curatorial space, a means for me to select and engage the types of stories and critical approaches necessary to reinvigorate conversations about OutKast's contributions to hip-hop. OutKast is the first southern hip-hop group to gauge contemporary scripts of blackness while referring to the past to annotate their southernness. "OutKasted Conversations" served as a digital complement to OutKast's undertaking of the continuous task to recognize southern black folks' cultural and sociopolitical agency in a more contemporary form. Perhaps most importantly, "OutKasted Conversations" experimented in digitizing the experiences of the contemporary black South, an effort to create a "playlist" of interviews that conceptualize and add depth to considerations about how hip-hop and regional identity merged to create new digital identities in the post–civil rights American South.

Aesthetic Influences

"OutKasted Conversations" is a critical dialogue series recorded on Google Hangout and hosted on my YouTube channel. The series concludes with a playlist boasting forty episodes and interviews with fans, scholars, and artists who enjoy and are familiar with OutKast's work. The interview and conversational format borrows from Mark Anthony Neal's "Left of Black" series of webcasts. Neal's use of social media as a platform for public scholarship and education is a useful model for connecting critical frameworks to nonacademic audiences. "Left of Black" features an interdisciplinary focus that provided context for crafting "OutKasted Conversations" as a site for multiple entry points of analysis about the contemporary South using OutKast's work. "OutKasted Conversations" uses new media as an intervention for new southern black studies using hip-hop. I used this project to extend conversations about the post–civil rights South offered by scholars like Imani Perry, Zandria Robinson, and Riche Richardson.[2] Perry, Richardson, and Robinson include OutKast in their analysis of southern identity politics and spaces, but they do not centralize the duo's work in their respective studies. "OutKasted Conversations" is the first project of its kind to centralize OutKast as a cultural framework for analyzing race and identity in the post–civil rights South.

Further, this project's social-media format evokes Zora Neale Hurston's approach to ethnographic study. Hurston's training as an anthropologist allowed her to document and record southern black folklife in the 1930s for the Works Progress Administration. Her influence is significant to this project because she was a black woman archiving southern black life while participating in the culture she observed. Myron Beasley's discussion of Hurston's place as a subject of digital scholarship (chapter 2 in this volume) adds further context to considerations of how black culture resonates within digital spaces. Beasley acknowledges Hurston's sonic ethnographic studies as immersive and necessarily self-subjective scholarship. Hurston's use of sound and recording tools created an alternative space for articulating southern blackness. It offered Hurston the opportunity not only to record the stories of marginalized southern black folks but also to record herself and her perspective into cultural history and memory. Beasley writes that Hurston's sonic work "eliminate[s] boundaries between the scholar and the participants and mak[es] known the cultural politics of doing fieldwork and producing creative and accessible ways of (re)presenting scholarship and creating new texts."

Additionally, Beasley's marking of digital media as comprising a "contested space" doubly binds the scholarly development of technology to region and gender. Hurston's sonic ethnography laid the groundwork for my own because it intentionally existed between the grooves of its audio recordings, purposely inhabiting the interstitial spaces between what is considered traditional and public scholarship. Hurtson's body of work recognizes that the (rural) black South did not fit onto a typecast page or within the framework of traditional critical anthropology. Thus, she used sound and sound production in all their manifestations—whether she was literally spelling out dialect in her creative writing or recording the sound of her own voice as it connected to the larger conversation taking place via southern folklore and song. Her body of work ruptures cemented expectations of race and scholarship in the academy and among the public. She demonstrated the highest levels of public scholarship by situating herself within the public. Hurston is a part of the culture she studied, which left room for her subjects to tell their own stories in their own ways.

As a southern black woman scholar working with digital media, I find Hurston's model of sonic ethnographic study useful for creating space to think through and record the experiences of those viewpoints otherwise overlooked in cultural studies. Each episode serves as a mini-rupture or intervention that leads to viewing OutKast (and ultimately southern black popular culture) as a framework for contemporary black identity. Like Hurston, I actively participated in each interview, sharing my own stories, humor, and excitement about the lasting relevance of OutKast's work on my self-identification as a young southern black woman. My engagement with each interviewee states my vested interest in their stories and ideas. My being "present" as a subject as well as the moderator of each conversation—just as Hurston was in her ethnographic studies—allows "OutKasted Conversations" to blur the lines between curating articulations of southern blackness and participating in the articulation of southern blackness. This is important because black cultural expression, especially southern black culture, ebbs and weaves between active participation in culture and its creation.

Interviewee Selection and Discussion Question Samples

To discuss the significance of OutKast's contribution to popular culture, I intentionally selected the majority of project interviewees for their southern backgrounds or intimate knowledge of the South. Their southern sensibilities came from multiple vantage points—I interviewed guests who grew up in Mississippi, Georgia, Louisiana, Florida, Tennessee, and Texas—which lends credence to the project's main objective: identifying and clarifying how OutKast signifies a complex and nonmonolithic southern black experience. Interviewees were also selected for their fresh insights, innovative scholarship, and willingness to help promote the series. Upon their agreement to participate, I sent each interviewee a list of questions to help steer the direction of each interview. Questions were geared toward the interviewees' area(s) of expertise. The resulting conversation led to a unique and exciting use OutKast's work to understand race and identity in the contemporary American South.

Crafting "OutKasted Conversations"

Each interview began with the question "How did you become OutKasted?" This question is pivotal to the entire interview. It is a unifying thread of commonality for the project and breaks ground for archiving one's personal experiences with OutKast. The question also speaks to the significance that the act of listening plays in articulating a cultural framework of one's lived experiences. As interviewees shared their stories they also revealed how they listened to OutKast, when they listened to OutKast, and why they listened to OutKast. Their responses laid the groundwork for more traditional methods of analysis to take place in the interview. The act of listening served as a primary method of engaging OutKast's music as a critical framework for race, class, and identity politics in the post–civil rights South. "OutKasted Conversations" collected stories about the varied listening practices surrounding OutKast's music. I used them to create a cultural reference point for contemporary southern black culture. The act of collective listening overlapped with the act of "collective watching" via YouTube. Both the series and the digital platform are grounded in personal tastes in streaming, forms of consciousness, listening preferences, and sociocultural attachments. The interviews extend the way collective cultural memory on a single subject can merge and "stream" in digital spaces. For example, in an interview with DJ Jelly, the first DJ in Atlanta to play OutKast's breakout track "Elevators" from the ATLiens album, Jelly discussed his initial listen of the song on vinyl. Jelly's discussion of breaking the record on air using a vinyl LP demonstrated the collective act of listening: radio listeners calling to request the song after hearing it, OutKast's transition from a local Atlanta hip-hop group to the national hip-hop stage, and the physical act of listening—which encompassed the transition from vinyl albums to compact discs and highlighted the role of the DJ as a curator of sonic cultural memory and experience. Asking interviewees about their initial experiences listening to OutKast positions listening as an act to collapse binaries of public/private cultural markers and gendered expressions of southern identity.

Further, consider episode four, which features Dr. Treva Lindsey, an assistant professor in the Department of Women's, Gender, and Sexuality Studies at Ohio State University. She is also a member of the Pleasure Ninjas Collective, a group of black feminist scholars who interrogate pleasure as a form of resistance and reclamation of power in black women's lives. Lindsey's episode focuses on the connections between the sonic and pleasure in OutKast's work. Lindsey's theorization of "user-friendly" patriarchy highlights the nonabrasive yet misogynistic undertones of women's narratives heard in OutKast's music while also pointing out how their sonic cues of womanhood—moaning and laughter, for example—demonstrate the rich complexity in utilizing OutKast's music as a critical framework for understanding gender and sexuality in hip-hop. Lindsey's interview offers sound as an alternative framework for analyzing contemporary issues of race and sexuality. The digital format of the interview was useful here because Lindsey could sonically demonstrate the oral indicators of black women's sexual politics used in OutKast's music. The video interview allowed for a sonic and academic performance of the Lindsey's analysis, offering an immediate and engaging critical insight into OutKast's work.

Each episode serves as a multilayered standing reservoir of contemporary scholarship. The topics addressed throughout the series—from gender and pleasure politics to automobile culture to film studies—work well in a digital platform because of its immediate access. Unlike a traditional print journal article, where the publication process can span anywhere between a year and five years, digital scholarship is immediate and can be immediately applied to cultural studies and discussions taking place in the present. Additionally, the immediacy of digital scholarship feeds into the fickleness of public interest. The "OutKasted Conversations" series took advantage of a sociohistorical moment when interest in OutKast—who have not released new music as a group in the last decade—reignited to celebrate their international twentieth-anniversary reunion tour in 2014. The public's interest in OutKast (including those who came of age on their music and those who only knew them because their Coachella performance raised curiosity about who they were) helped buoy the progression of the series throughout its production.

Further, the dialogue series signifies the blurring of the academy as a private and publically unresponsive space. As I state previously, the intimacy and lightheartedness of the conversations emphasize the crossover appeal of a cultural subject like OutKast in both academic and lay spaces. Viewers have access not only to the academic discourse but also to the voices behind the analysis. The interview documents not only the analysis but also its delivery. The critical engagement is not lost but reimagined to speak to a wider audience than exists inside the classroom or between the pages of an academic journal.

Process Editing

After the conversation was recorded, the raw footage was downloaded and edited with video software (iMovie). I minimized editing to preserve the organic flow of the conversation and to keep intact the critical work being done. Perhaps the most beneficial aspect of the project for me was undergoing a public version of peer review for my work. Rather than relying on academic experts in the field to offer insight, I relied heavily on my viewership to help me improve the format and functionality of the project. Feedback was quick, personable, and utilized with a quick turnaround in the project's production. For example, the earlier interviews of "OutKasted Conversations" (episodes 1–10) are minimally edited video from a conversation recorded on Google Hangout with an attached title slide. Episodes were long, ranging from thirty-five to sixty minutes.

After receiving feedback from viewers and consulting new media strategists like Mark Anthony Neal, Marisa Parham, and my partner, Roy Bradley, I sought to make the episodes more polished and to retain audiences by cutting down the length of each episode. I switched the format to include a title slide, an introduction slide listing the guest's name, and end credits. Each episode only lasted from fifteen to twenty minutes. Starting with episode 11, a friend and music producer, J. French, gave me an instrumental track to use as the series' theme song. The song played approximately five to seven seconds and faded out after the slide introducing the episode number and name of the guest. To further polish the final product, I added a photograph of the featured guest to the introduction slide. I then exported the segments from iMovie and uploaded the final product on YouTube. Uploading episodes on YouTube made me stick to a weekly production schedule—filming the episode and editing it a week in advance of its airing—to keep drumming up viewer interest and maintain a consistent presence on social media.

Publication and Advertising

The polished segments were uploaded weekly to YouTube and shared via Twitter and Facebook. I would tweet the link to the project using the hashtag #OutKastedConversations to track its movement across social media. I also tagged OutKast member Big Boi to alert him to the series and new episodes. By advertising via social media, I hoped to achieve additional conversations about the episode and OutKast with a broader scope and audience. Indeed, I achieved a broad audience. "OutKasted Conversations" realized nearly eight thousand unique hits, and over two hundred users subscribed to my YouTube channel. It was featured in major digital media publications like *For Harriet*, *Sounding Out!*, *Creative Loafing Atlanta*, *Huffington Post Live*, the *New York Times* popular culture blog, and the *Feminist Wire*. The project garnered fanfare on social media in the form of retweets, direct mentions, and Facebook (re)posts. Although there was significant support from public platforms, there were few fan emails or correspondence outside of the publications previously mentioned.

Lasting Impact

"OutKasted Conversations" stands as a public archive of southern hip-hop collective memory. I am currently in conversations to move it to a more stable digital platform. The focus on OutKast serves as intervention to include more southern voices—both literal and conceptual—in the canon of southern studies and hip-hop scholarship. Social media provided me a platform to engage a subject matter and explore perspectives otherwise overlooked in the academy. "OutKasted Conversations" exists at the crux of sound studies and new southern black studies because it interrogates how critical voices and expertise legitimize themselves outside of academic discourse. Like the black southern oral traditions studied and documented by Zora Neale Hurston, "OutKasted Conversations" became a space of collective reckoning about how the South is rendered from a post–civil rights southern black perspective. OutKast served as a subject and as a springboard for renegotiating contemporary black agency for those generations removed from the historical civil rights era. These types of conversations take place in cars, around lunchroom tables, or through phone calls and texts. Public discussion can overlap with academic study to create new discourses and add deeper contexts. "OutKasted Conversations" reflects the overlap of popular and academic study by using alternative methods of analysis like sound and social media. It is a testament to the multiple possibilities of using hip-hop culture in digital spaces to update the South to reflect its present and future states.

NOTES

1. For links to all forty-two conversations, please visit my website at www.redclayscholar.com.
2. See Perry, *Prophets of the Hood*; Robinson, *This Ain't Chicago*; and Richardson, *Black Masculinity*. These studies contextualize OutKast and their scripts of blackness and masculinity within the framework of a contemporary and urban/postindustrial South.

WORKS CITED

Neal, Mark Anthony. "Left of Black" webcast series. Accessed November 12, 2107. http://leftofblack.tumblr.com.

Perry, Imani. *Prophets of the Hood: Politics and Poetics in Hip-Hop*. Durham, NC: Duke University Press, 2004.

Richardson, Riché. *Black Masculinity and the U.S. South: From Uncle Tom to Gangsta*. Athens: University of Georgia Press, 2007.

Robinson, Zandria F. *This Ain't Chicago: Race, Class, and Regional Identity in the Post-Soul South*. Chapel Hill: University of North Carolina Press, 2014.

REPROGRAMMING SOUNDS OF LEARNING

Pedagogical Experiments with Critical Making and Community-Based Ethnography

W. F. UMI HSU

Teaching and learning are a series of interpretive acts. From designing a syllabus to enacting classroom exercises, teachers construct the value of education by assigning outcomes of learning to grade values. Students maintain the value of education by performing tasks in order to achieve the goals of classroom activities and assignments. These processes resonate with programmatic acts such as encoding, decoding, and enumerating. In many ways, pedagogical design is very similar to software design. Computational logic pervades much of the thinking familiar to teachers and administrators. For instance, at the curricular level, programming means breaking down the experience of learning into uniform components and then counting, sorting, and grouping these components based on the mission and the objectives of a degree program. At the course level, grading exemplifies a *markup* activity that ranks student work; and *scripting* in-class activities sequences interactions and governs informational flow among the students and the instructor.

Codifying learning leads to the evaluation of the learning outcomes across metric categories that have been standardized. This process of codification interfaces with the myriad modes of learning, from reading and writing to classroom discussion and testing. These standardizing practices rank modalities of learning based on a hierarchy of senses that prioritizes some experiences of learning over others. For example, class participation is typically an embodied experience—including raising hands, voicing an inquiry, exchanging ideas with peers during a class discussion. In evaluation, the metric of "class participation" has lumped these sonic modes of learning into a single category. This category is often ascribed with little weight relative to other categories based in silent modes of learning such as final essays, midterm exams, and reading responses. The ordering of senses results in the privileging of writing and printed text over auditory processes such as listening, speaking, discussing, making, and collective brainstorming.

Given the compulsory silencing of institutional learning, I ask: How would a sounded pedagogy reorganize the communications and information flow in learning? What might be some guiding principles for thinking about a sound-based approach to teaching and learning? How might "sonifying" learning encourage students to explore a personal meaning of learning? Can sounds enable students to encode and decode knowledge reflexively across various contexts of learning? If so, how? Finally, how does a sound-based pedagogical approach foster collaboration and community building? In this chapter, I first offer a critical perspective on industrial models of pedagogical designs and practices that encode sounds (out) of the learning experience. Then I propose a series of experimental approaches that attempt to reprogram sounds back into learning and teaching.

The Code of Silence

Looking at syllabi from the past, I noticed something unusual on the History of Civilizations syllabus for a course offered in 1969 at Occidental College. In a description of the journaling assignment, the instructors state that the journal is "not a place to sound off."[1] The use of the term "sound off" struck me as a peculiar way to refer to complaints about professors. Sounding off typically involves speaking loudly, an act that comes with a distinctively audible component. The anti-sounding-off restriction on the syllabus poses an unexpected dissonance to the assignment of journal writing. Journal writing is typically an internal, individualized grappling with intellectual materials. Placing a restriction on a sounded speech act within a quiet, introspective writing exercise seems out of place. How did the instructors of the course imagine the sounds of learning? Are they necessarily associated with unruly classroom behaviors? Did they see the need to exert control over sounds so badly that they had to extend their policy into the sphere of individual journal assignment?

In higher education accepted modes of learning, reading, and writing are traditionally associated with quietude.[2] Libraries, with quiet floors and individual study carrels, are conventionally designed as spaces of silence. Even processes of learning academic subjects with an aural emphasis like foreign languages and music are contained within and isolated by laboratories equipped with individual stations with headphones. The silencing of learning extends into the course design. Courses in the humanities and humanistic social sciences are programmed by a series of readings and evaluations. Students reflect on their learning by quietly writing a final paper and testing their knowledge in an exam.

Working individually and silently makes students submit to authority. It can also suppress student impulse to question the purpose and modality of education. These classroom designs and course policies are aligned with the industrial mission of training students as good, quiet workers. The quiet worker evokes Paulo Freire's diagnosis of education as a *banking model* that operates as a bureaucracy to maintain order and promote efficiency. In this industrial model, students and factory workers are objects that can be quantified for the purpose of resource and labor management.[3] This metrication, the process of turning the human experience of learning into metrics that evaluate student performance, is in place to increase productivity and efficiency. In some instances, metrics are implemented to quantify faculty salary and other resources that go into the delivery of a class. The purpose of metrication, in the capitalist-industrial context, is to drive growth. For teachers this means the imperative to increase course enrollment, and for students the objective is to obtain higher grades in order to compete on the job market postgraduation. The grade-driven incentive for student achievement reinforces the data-driven paradigm of classroom management. The continuity between the data- and grade-driven paradigms flattens the purpose of education by producing an efficient, compliant workforce. It is worth asking what other learning objectives are important besides training students to become productive workers without an inclination to sound off, especially in the current postindustrial economy.

Furthermore, the capitalist-industrial logic of course design can reduce the richness of learning to a binary between sound and silence. In its most simplistic case, it turns the sounds of learning activities on or off, like a switch that allows for silent activities such as writing and reading. In other instances, the industrial logic enforces the transmission of sound in a single direction, with the classic paradigm of a professor lecturing over a crowd of silent note-takers. These programmatic mappings in learning design often privilege silence over sound, writing over speech, reading over discussion, thus reinforcing the instructor's authority over participation and interaction.[4] What if the experience of learning could resound in a full spectrum between sound and silence, including noise, music, whispers, provocation, recitation, call and response, and other relevant sounding experiences? A reprogramming of learning calls for the rethinking of the role of sounds in learning beyond the dichotomy of sound and silence, ushering in classroom dynamics with sounds and noises that emanate from the bottom up, sideways, and across.

Reprogramming Sounds

Sounds can chart new territories of learning. They can amplify the tacit and reembody a message, a set of instructions, and a corpus of knowledge. They can renew textures of knowledge, bringing into existence interpretations and inquiries of personal and social significance on a journey of learning. A sonic rehabilitation of learning can remodel the mission of education and reconfigure pedagogical relationships. Nuances of sonic modality and mediation are central to the process of acquiring and embodying knowledge. Sonically informed insights can give us ideas for creating engaged learning. Foregrounding sound as a medium and modality of learning, I want to draw attention to how sound dynamically registers at the experiential, ideological, and societal levels.

I advocate for a pedagogy that encodes sounds into the learning scaffold. This act of reprogramming begins with raising sound-first inquiries about teaching and learning so that sound is a central principle and not an afterthought. I employ examples from Digital Music-Cultures, a course I designed and taught in spring 2013 while experimenting with digital pedagogy and multimodality with and through sound. Combining principles of ethnomusicology and digital audio production practices, Digital Music-Cultures is an entry-level music course for nonmajors.[5] To create a new pedagogical schema, I identified points of intervention that could be meaningfully sonified through rescripting class discussion, workshops, homework assignments, and final projects.

It is worth noting that even though the subject matter of this course is music, a sonic medium, most nonperformance-based music courses, such as music history and ethnomusicology, are taught in ways that are confining sonically. For instance, the listening portion in a similar course usually manifests either as a take-home assignment for individual students to engage with privately, in their own time, or as a drop-of-the-needle identification portion of a written test. A pedagogical goal of this course is to reorganize the experience of sounding and listening so that they are central to learning. Like a choir rehearsal, evoking sonically driven learning practices such as a call-and-response ideation, a performative demonstration of feedback, is treated as foundational to the course experience.

I propose three principles to reprogram the way sounds are learned: remediation, reflexivity, and resonance. These principles are derived from a series of pedagogical experiments I conducted while teaching undergraduates from 2006 to 2013. All three interrelated and non–mutually exclusive principles demonstrate the intersecting possibilities between sound as a medium and the digital as a modality. In what follows, I elaborate on each of the enlisted principles with actual examples drawn from the course.

Remediation

Remediation refers to the transfer of content in one medium context to another.[6] Remediation occurs as content becomes represented across media contexts: for example, a film adaptation of a theater production, song lyrics derived from poetry, or photographs of paintings. The concept can also be exemplified when content transfer happens across format types: from analog to digital, from radio show to podcasts, from vinyl recordings to MP3 files.[7] Further, in digital humanities, remediation can be theorized from the perspective of materiality. Challenging assertions of digital immateriality, digital humanities scholars have conceptualized the materiality of digital objects and processes.[8] Digital affordances, they contend, enable knowledge transformations. Through these transformations, objects of knowledge are reiterated and reembodied across modalities and media types.[9] The technological possibilities for visualizing textual and sonic materials have enabled humanities scholars to manipulate the form and format of cultural content, renewing the analytical context for discovery and insights.[10]

The pedagogical value of remediation or rematerality becomes evident when students wrestle with often-challenging intellectual processes across media contexts. Reading, discussing, making, listening, sharing, rereading, remaking, relistening, rewriting—these tasks are iterative remediations of concepts and theories from a course. Each time students remediate course materials, from reading to writing to discussing, they develop a deeper and more nuanced relationship to those concepts.

Sonification—the act of turning nonsonic materials into sounds—is well poised as a remediation practice for providing a new sensory context for students to grapple with knowledge. It is a space for students to articulate relational knowledge: for instance, exploring the relationship between their own argumentative positions and sources of scholarly materials. Sonic remediation of student writing, in particular, can help students hone their arguments with respect to other scholarly voices and content. In my writing-oriented courses, I always structure an assignment asking students to record themselves reading a previously composed essay of their own.[11] This assignment allows students to explore their authorial voice within the sensory domain, enriching the experience of writing. I often see students attempting to sound "scholarly" in their writing. This exercise disabuses them of notions of having to sound scholarly. Instead of sounding like a generic scholar (whatever that means in their heads), I want them to take control, to reclaim their own voices, and to embody argumentative writing on their own terms. Sounds can also afford us opportunities to remediate scholarly concepts, which are almost always transmitted as printed text. In this instance, sonification can be an interpretive exercise that provokes recontextualizations of meanings and knowledge. Using sounds to rematerialize scholarly information, students can gain multiple access points, including those that are embodied, sensory, and potentially affective, to enter into the scholarly conversation and develop a personally meaningful relationship with the object of intellectual inquiry.

To experiment with sonic remediation, I ask students to sonify their responses to their reading of a theoretical text and their viewing experiences of a documentary film. For a unit on chip music, I created an in-class exercise for students to explore concepts of music and noise described in an article excerpted from Jacques Attali's book *Noise: The Political Economy of Music* and in Paul Owens's documentary film about chip music, *Blip Festival: Reformat the Planet*. First, students work in groups to populate a shared Google doc with quotations from the Attali article that address concepts related to music and noise. I scaffold this class activity by providing prompts to evoke possible theoretical engagements.[12] Students populate a list of Attali quotations and annotate each quotation with an analysis of how the quotation offers a perspective on chip music as depicted in the film.[13] In the following class meeting, held in a media production workshop, students acquire the basic techniques of chip music production, learning to compose music in Little Sound DJ, a beat-making Game Boy game simulator. For their take-home assignment, students compose a chip music piece as an audio meditation on theories related to music and noise, while referring to the peer-sourced list of quotations from the previous class meeting. In a reflective blog post describing their results, students discuss how their composition does one of the following:

- exemplifies or reflects an ideology (related to music, society, consumption, or technology) expressed by Attali,
- demonstrates a technological or musical concept discussed by Attali,
- contradicts how music (or noise) is defined by Attali, or
- explains or encapsulates the meaning of music (or noise).

This creative assignment encourages students to engage with sonic argumentation—to demonstrate, extend, or undermine concepts in the reading—through audio production techniques.[14] This multipart lesson ends up creating a space for students to speculate on the triangulation between three learning components of the unit: high theory authored by a canonical scholar, the grassroots community of chip music practitioners depicted in the film, and the practice of chip music audio production. Sound, in this example, serves as a remediating agent that grapples with the relationship between two texts in two different media, across two interpretive domains.

Possible intellectual productivity comes to life when the students' deployment of an aesthetic decision via audio software techniques interlocks with their explorations with scholarly concepts. Something *clicks*—an experience we have all had in learning—and the fruits of interpretive efforts emerge. Interestingly, "clicking" is an auditory expression of a productive moment of intellectual grasping or knowledge discovery. A famous example of this is Archimedes's exclamation, "Eureka!" I wonder if the recovery of the sonic dimension in learning could spur meaningful "eureka" moments.

This assignment encourages students to engage with a deformative path to imagine new and creative forms of scholarship that can be "forbidden . . . either irresponsible or damaging to critical seriousness."[15] Linking deformative reading with digital making, Mark Sample champions a making approach rooted in breaking things as a predominate mode of making new cultural objects.[16] A deformative making project refuses a "revitalized perspective," deliberately not treating a new text or artifact as a derivative or secondary object in relation to the original text.[17] This lesson on noise and chip music itself is modeled after the deformative, hack-based praxis rooted in the chip music and related noise music communities. Parsing and breaking Attali's text into creative scraps with the potential to germinate new systems, I believe, is a deformative act. Multiple students play with an Attali quote that articulates a historical homology between music and technology: "Every code of music is rooted in the ideologies and technologies of its age, and at the same time produces them."[18] Some students find a way to engage with Attali's writing in their chip music composition. A few students recompose popular tunes from their own time (c. 2013) using sounds produced by the Game Boy chip music emulator to reflect the idea of age, to show the temporal disparity between the technological relic of the Game Boy and the tech of their present. Other students use even more abstract parts of Attali's text to sonify the idea of noise as a means of materializing the relationship between music and human perceptions of chaos and noise. Deformative approaches to pedagogy have even greater implications for rethinking the role of creative assignments in humanities courses. I will reflect on this corollary in the final section of this chapter.

Reflexivity

Reflexivity describes a system that refers back to itself. It models the feedback loop and embodies circularity. The concept of reflexivity has implications in music studies, digital humanities, and media studies.[19] I offer ethnomusicological insights on the relationship between reflexivity and the transmission of knowledge. In particular, I draw on Tomie Hahn's work that looks at the transmission process of embodied knowledge in *nihon buyo*, Japanese traditional dance. She declares the critical positionality of her personal experience in her monograph: "Because *nihon buyo* has been a part of my life since childhood, it was a clear candidate for a case study on the transmission of cultural knowledge. I decided to write this ethnography with a reflexive voice because my body physically experiences and informs my perspective on transmission, and ignoring this voice would have been disingenuous."[20] According to Hahn, the process of knowledge transmission is central to ethnography. The researcher's reflexive forms of knowledge can be critical to grasping cultural knowledge through wrestling with the tension between self and other and through embodied, tacit ways of knowing.

Learning is, in many ways, an exercise of research, a process of knowledge discovery and transmission. The reflexive framework offers a fruitful perspective regarding the purpose of learning. In institutional learning students often take for granted the value of learning. The product orientation of learning becomes a barrier for students to realize the transformation potentials of knowledge. Learning something for oneself begins with the realization that the process of knowledge acquisition can be personalized. Learning can be a process of self-becoming, and knowledge acquisition is not an end goal but a process that can be meaningful in itself. Self-knowledge, as Hahn reminds us, can be a "resource within research."[21] How can we reposition learning as something that's more process-oriented? How can we rearticulate the purpose of learning? My answer to this question is a reflexive ethnographic final project that echoes Hahn's ethnographic research framework.

Dubbed Sounds of Learning, the culminating class assignment is a community project that pairs college students with sixth graders from a nearby elementary school to coproduce a three-minute audio piece that documents and comments on youths' experiences of school and learning. Based on a reflexive logic of learning about learning, this project extends classroom learning beyond the confines of a college to embrace broader notions of cultural and embodied learning in the community. Using a community-based, ethnographic paradigm, this project recodes learning by embedding students in sounded communities. The project explores, activates, and records the sonic dimensions of acquiring, mastering, and embodying new information, cultural knowledge (pop culture, heritage, language), social norms, and values (identity, status). I introduce this collaborative provocation using the text below:

> We deliberately sound the process of learning by asking our sixth-grade collaborators to capture sounds that are meaningful to them. These might include the sounds of school activities and environment, conversations with peers and adults, interactions with popular culture and media, sounds of home and neighborhood, and counting. In the most literal sense, the sounds of your interview with students of Annandale Elementary—what and how they articulate as their answers to your prompt questions—are sounds of learning in themselves. They reflect

how the sixth graders come into awareness of their surroundings. As importantly, these sounds teach you, the ethnographer, aspects of the social and aesthetic world that they live in.[22]

During this collaboration, students enrolled in my course synthesize appropriate techniques and ethics of ethnographic research and field recording that they acquired throughout the semester. Through a hands-on engagement, students reinforce their knowledge of another course premise, ethnography as an embedded and sounded practice.

In this model, recording is considered as a reflexive research practice that extends the technique and purpose of close listening. "Recording is itself a form of research. Of course it is important for a documentary producer to capture good sound, but getting any kind of recording is also a mode of exploration and investigation in its own right."[23] In this project, I challenge students to think beyond the expected content, form, and standards of recording quality of "sounds of learning." Students should continue to reexamine their definition of learning, throughout the progression of the final project by working through destabilized notions of aesthetic worthiness and acceptability while interpreting field recordings.

The digital making component of the Sounds of Learning class project evokes some of the technique that others refer to as critical making. Matthew Ratto ties critical making to the mission of synthesizing theoretical and pragmatic modes of engagement with knowledge that is often held separate: "Critical thinking, typically understood as conceptually and linguistically based," joins with "physical 'making,' goal-based material work."[24] Deconstructing the recipe of how digital sound media are made via an act of remaking can afford students of both Occidental College and the partnered elementary school to gain an access to personal and reflexive meanings of technology in their everyday lives. This kind of critical making can also help recontextualize students' relationship to technology, enabling them to question their expected role as technology consumers and end-users and engage with technology beyond the black box.[25]

Resonance

Sounding and listening are both relational and social activities. They bridge social rifts and forge new connections. They generate resonance and social openness.[26] In a learning context, sounds can facilitate and encourage the exchange of information between multiple sounding agents, for example, between the instructor and students, between students and their classmates, between students and their extended peer and family networks. In this resonance framework, sounds can activate participatory learning and empower individuals with a voice to express themselves. Allowance of sounds and voices can flatten the social hierarchy of the agents in a classroom. Jesse Stommel shares Freire's vision of "problem-posing education" as an alternative to the industrial banking model of education: "A classroom or learning environment becomes a space for asking questions—a space of cognition not information. Vertical (or hierarchical) relationships give way to more playful ones, in which students and teachers co-author together the parameters for their individual and collective learning."[27]

Sounds can be a medium of power for individuals to assert their agency. They enable the activation and emanation of voices, an articulation of difference and plurality that can be heard by the participants and their audiences. A vocal enactment of plurality can undermine conventional classroom dynamics and redefine the purpose of education. In what follows, I will draw from the Sounds of Learning final project to illustrate the affordance of sound as a catalyst to reorganize the traditional flow of communications related to teaching and learning.

Occidental College straddles two neighborhoods in northeast Los Angeles: Eagle Rock and Highland Park. Though with slightly different social histories, both neighborhoods have been changing dramatically in terms of land and property values. The community discussion about gentrification and displacement (the dispelling of low-income renter-residents in the previously predominately Latino neighborhoods) has become more polemical. My course took place in 2013, a time when signs began to show of neighborhood changes related to real estate and property development. From informal conversations with community organizers and the director of the college's Center for Community-Based Learning, I gathered that a partnership with the local elementary school would not be seen as politically neutral. Occidental College students have traditionally been uninvolved in activities in the broader community. This "campus bubble" and the social divide between the Occidental campus and the broader community is perceived as a reality by the community and, to an extent, by the students and faculty themselves.

With a goal to create a shared experience based in colearning, I set out to intervene in the existing power relations between student participants at Occidental College and at Annandale Elementary School. In addition to the age disparity between the college students and the sixth graders, other social factors mediate the ideas of difference between these two groups. Occidental is private liberal arts college. My students were mostly white, whereas the Annandale sixth graders were majority Latinx. While a subset of the Occidental students are first-generation college students, some of whom are on financial aid, the majority of the student body consists of students who come from socioeconomically privileged backgrounds. This project provided a platform for my students to conduct community-based research with an emphasis on researching *with* a community—in other words, observing and participating in the social lives of their sixth-grade research partners.

This collective research model privileges the experiences and the epistemology of the sixth graders, thus making the Occidental students assume the role of learners of the social world in which their elementary-school partners live. Sound is foregrounded as the medium of this unique learning journey while engaging with processes of knowledge transmission, speculation, and argumentation. At the kickoff meeting, which takes place at the elementary school, college students meet and teach their sixth-grade partners the basic techniques of field recording. Sixth-grade students then take recorders home with them with the goal of gathering recordings related to learning. During the field-recording period, a workshop is set up for sixth graders to share their recordings with their college student partners. They work together to coexplore the meanings of the recordings; based on the outcome of this exercise, they may restrategize their field-recording plans. Then college students meet during their class time to review ethical principles of ethnography and develop a set of interview guidelines.[28] In a final digital-making workshop, college and sixth-grade students discuss, negotiate, and eventually come to an agreement on a shared production vision and plans for the final composition. Following their agreed plan, college students spend the final two weeks of the semester listening closely to the recorded materials while mixing and editing recordings into a composition. A listening party takes place inside the elementary school's multipurpose auditorium at the end of the semester, bringing together sixth-grade students, teachers, family, college students, and administrative support staff of Annandale Elementary and the community-based learning center at Occidental College. Based on the feedback gathered at the listening party, college students revise the mixes and submit their final version along with a thousand-word blog post reflecting on the project in light of concepts learned throughout the semester. The final mixes of the compositions are distributed back to the elementary school students with cover art and liner notes created by one of the Occidental students.[29]

For my students, the success of their projects is highly dependent on the recording products collected by their sixth-grade partners. Throughout my students' workshop, I hear students complain that their sixth-grade partners did not collect adequate recordings. One student claims that his sixth-grade partner's recordings of neighborhood streets are meaningless in the context of the assignment. I take these complaints as opportunities to push my students to listen harder and think critically about their own assumptions about these sounds and their partners. I pose questions such as: Do you hear sounds like this in the neighborhood where you grew up? If not, how do you make sense of this difference given what you know about their social world? These inquiries ultimately lead students to interrogate their processes of knowledge production and assumption formation within the project's social specifics and to expand what they consider to be legitimate knowledge.

Throughout the project, my students are encouraged to form a dialectical relationship with the sound of learning accomplished by, to use a metaphor introduced earlier in this chapter, the encoding and decoding of culture. The acts of encoding and decoding cultural materials—specifically, making, remixing, and composing with field recordings—constitute the core of learning, the acquisition of knowledge. Through recording, mixing, and composing, students listen thoughtfully and kinesthetically across barriers of education, class, age, gender, and ethnicity. And the bidirectional relationship with research associates achieved through shared listening and making help cultivate empathy, a desirable quality that emerges from reflections of ethics and critical positionality in ethnographic research. One student articulates this outcome in his final reflection essay:

> My partner, Anthony, was very humble at first, but later opened up to me during his visit to Occidental College. I did not understand the world he inhabits as he described when we first met. While he did his best to paint a picture in my head, I could not get a clear image without a sonic environment. It was only after I listened to his recordings that I was able to visualize a picture of his world. . . . I originally thought of a school setting, but through my interview with Anthony, I began to think with a wider perspective and settled on focusing on his life at home. Learning is not only math and science, but also life lessons and growing up. An-

thony shared with me that his parents were divorced and he did not really have a place where he can call a permanent residence. Home is a place of learning because that is where someone grows up and develops personality. Even though Anthony lives in different homes at different times with mom and dad, he felt that both places were his home. I was really grateful that Anthony was able to open up to me and talk about his family. I made a lot of effort to engage in casual conversation to make him feel comfortable to just talk story and not pay attention to the recorder. In the end, he told me jokes about pranks that his family members did to each other. He even shared that his dream is to become a Marine just like all the men in his family. He asked me questions about what college life is like and other things that are not particularly relevant to the music project. Just like John and Alan Lomax were able to do field recordings across the country by engaging in conversation and being friendly, I was able to do the same. The Annadale project taught me a lot more than I expected about myself and opened my ears to perceive a sonic world.[30]

This student's reflective excerpt begins with an assumption that learning takes place in school, but through conversations with his partner Anthony and listening to his partner's recordings, he is able to theorize more broadly about the meaning of social learning in his partner's life. The student author (who self identifies as idanxfi) is an international student from Japan. Listening to his Latino research partner across the ethnic and national lines ended up being a lesson about his own ethnic difference in relation to his partner. His reference to John and Alan Lomax hints at a deeper interrogation of the racialized relationship and economics of exchange between those who recorded (white, Anglo-Saxon) and those who were recorded (nonwhite, often black and Hispanic) in the history of folklore, an ongoing conversation throughout the term of the course. This excerpt illustrates that collective listening and digital making constitute a shared communication platform. Using this platform, student researchers may iterate the research cycle of listening to, speculating about, and making the meaning of sound while interviewing their research partners until they attain a deepened understanding of culture. This model exemplifies the multimodality of learning by sonifying the often inaudible learning spaces and processes by breaking down place-based conventions of learning, in this case school vs. home.

Colistening and comaking also question the subject-object binary that dates back to historical colonial research practices. "The Annadale project taught me a lot more than I expected about myself and opened my ears to perceive a sonic world."³¹ This has particular ramifications for Occidental College, a small, elite liberal arts college tucked in a semiurban pocket of metropolitan Los Angeles. The collaborative media-making process fulfills the mission of learning as a form of community engagement. Reducing the distance between subject and object of ethnographic research, critical and collaborative making encourages listening with empathy and communicating across differences.

For the sixth graders, this project serves as more than a technical arts workshop. It is intended to spark reflection and empowerment on a newfound understanding of their everyday cultural and environmental soundscapes, a discovery about how sounded environments have shaped their sense of place and self throughout their elementary school years. To this point about self-realization, CJ's project with Jazmine comes to mind. Over initial interactions, CJ learns that Jazmine is shy and uncomfortable with recording her voice. Turning this obstacle into an opportunity to forge a connection, CJ repositions his role relative to his research partner. He recounts this moment in his reflective blog post: "This project became more than just a way to get a good grade in the class, but rather an opportunity to shape someone's life. My role moved away from mentor, interviewer, and ethnographer into cheerleader, motivator, and empowerer."³² Jazmine is an aspiring singer but refuses to sing in the presence of CJ. As a response, CJ, who is also a singer, encourages Jazmine to explore her voice through self-recording and operate the recorder herself as a means to take control of her own recording. CJ writes in his reflection paper:

> She became a different person, and the recorder transformed from merely a sound-capturing tool to a microphone of a singer. The recorder became a tool of transformation, a means to express identity, a source of empowerment. Though I had to walk away, once she was in her zen moment and alone, she allowed her soul to sing. The transformation was amazing.³³

CJ is highly aware of the representational politics of field recording and sampling, a topic of class discussions earlier in the semester. His relinquishing the control over recording shifts the typical dynamic of an ethnographic relationship. Teaching while empowering his partner Jazmine to record her own voice ends up bolstering the research associate's courage to take agency in staking a claim to her own representation.

In this instance, recording acts as an empowerment tool that disrupts the colonial and historical object-based thinking about documenting the cultural other. Recording has been reclaimed by the ethnographic subject, who not only uses technology to amplify her own voice but also acquires a transformed perspective about herself and her relationship to her own voice and embodied subjectivity. As a resonating medium between the ethnographic researcher and the research associate, sound enables the transformation of the directionality of knowledge transmission. It offers opportunities for both CJ and Jazmine to experience reflexive learning on their own terms and to play an active role in knowledge creation.

After Jazmine comes back with her recordings, she and CJ agree to make a "cool remix." With this self-critical awareness, CJ creates an audio narrative entitled "Blooming Flower." Interweaving the story with Jazmine's recordings of her own vocal explorations, CJ experiments with audio storytelling techniques that portray his partner's "finding identity and power."[34] The process and product of CJ's audio work support the affective and intellectual growth of his research partner's life. His thoughtfulness leads him to nuance a multimodal argumentation style that simultaneously critiques the medium and politics of ethnographic representation and builds a relationship with his research partner.

Final Reflection

Sounds are messy. They travel, leak, and cut through barriers that are ostensibly prohibitive. This makes sound a great medium to discover new paths for intellectual inquiry and practice. Sounds create opportunities to interrupt the existing logics in institutional learning. When I teach, I use these creative opportunities to reencode the meaning of learning. The most successful instances all call into question the product-focused approaches to learning. These teaching experiments require that I partially relinquish my control as an instructor to define what's meaningful in students' learning experience. Letting go of this impulse to script the purpose of learning means giving students their agency to determine their own relationships to their objects of inquiry. It also means that instead of meaning and purpose, I provide them with a scaffold to explore the web of scholarly knowledge with their own voices and positions. Reprogramming pedagogy in many ways means unprogramming some of the top-down command by the instructor. A thoughtful rescripting brings to life a dynamic learning algorithm that is reflexive, process-oriented, and participatory with student input.

Much of my effort in redesigning teaching goes into reconfiguring the relationship between reading and writing, listening and reflecting. "Breaking" text-based traditions in learning and using the deformative metaphor through sonifying can not only destabilize textual knowledge but also lead to unexpected learning results.[35] These sonifying interventions enable students to play and experiment with intellectual materials within a new space, one in which creativity plays a central role in learning. My intention for students to construct a sounded document—a new sonic artifact that by its existence has little respect for the original scholarly text—is to disrupt the fun vs. serious binary in university learning. Often in a university classroom, creative projects are relegated to a secondary place, treated as "fun" for extra credit or as a supplement to a more serious assignment like an essay. To a large extent the rigor of evaluating these creative works is underdeveloped in humanities courses because of the myth that creative work has no relationship with "serious" scholarly materials. My current articulation of the relevance of creative projects in humanities coursework hopefully contributes to the larger pedagogical conversation about the educational value in having students engage with multiple modalities of learning as they grapple with sometimes complex and esoteric scholarly content.

A few of these teaching experiments fail, however, as experiments do sometimes. One reason is students' lack of openness to try something different. Some students in my class had a hard time thinking outside the box. Many of them are first-year college students with habits of learning established in secondary education. Only a subset of students understood, for instance, the instructions provided for the chip music assignment and found a way to link their reading responses to their compositions. This could be because students are not used to engaging in creative practice in humanities courses and are often discouraged from tempering learning with subjective meanings such as affect, stories, and creativity. Within the course context, it may be useful to demonstrate this assignment by eliciting examples that explicate a link between intellectual and creative grappling with text, and doing so within the context of a transparent grading rubric.

While it is easy to assume that some students are "naturally" more creative than others, we as instructors need to be mindful of the effects of nontraditional learning engagements on students of various backgrounds. I notice that students from less privileged backgrounds are less likely to engage with learning experiments. What seems to me like a healthy challenge could end up being perceived as a stigma or being negatively tied to previous experiences of learning. Teaching interventions should be implemented with sensitivity toward the cultural and social needs of students and responsiveness toward the history of learning that each student brings.

Last, I want to return to CJ's story to offer a final comment. In his final reflection essay, CJ notes a resounding contradiction between the evaluative and transformative aspects of learning. "This project became more than just a way to get a good grade in the class, but rather an opportunity to shape someone's life."[36] How do we reconcile between the holistic mission of sounded learning and the competition-ridden assessment requirement of education?

Earlier in the chapter I spoke of how grading and evaluation rank the performance of student work. It seems strange that I critiqued this pedagogical practice but then do not examine it in the rest of the essay. While students and instructors may experiment with the meaning, media, and modality of learning within the context of a course, most of these practices are still fixed within the larger grade-oriented gridlock. While I apply the concept of resonance to reorganize the communication flow between student researchers and their ethnographic research associates, my relationship with my students remains status quo. I offer workshops and invite guest speakers into the course, but ultimately I still run the show. It is *my* course and I am still the authority as the instructor. What a conundrum. If given the opportunity, I would extend my reprogramming efforts into the realm of evaluation by considering alternatives such as a "contract grading" policy to offset the compulsory "transformation of a complicated, nuanced, and (ideally) supportive relationship into a mercenary transaction."[37]

The experiments that I have evoked in this chapter give fodder for thought. I hope to inspire further experimentation and iterations that come with the sharing of practices and the embracing of failures. Efforts to reprogram teaching and learning should happen at both the course and curricular levels. For a sustained impact, let us continue to imagine thoughtful and creative efforts that sonify acts of knowing and resonate transformative visions of learning.

NOTES

1. Winter et al., History of Civilizations.
2. The sonic politics of learning have class implications in earlier formations of social stratifications in the United States. Citing Cavicchi, Silva-Ford links the hierarchies of sound and silence encoded in learning to nineteenth-century ideologies that define class distinctions. Quiet behaviors of learning and read-

ing exemplify middle-class respectability. The quietude of reading and listening elevates the status of genteel people, setting them apart from the noisy pastimes of slaves, immigrants, and workers. Cavicchi, *Listening and Longing*, 52, cited by Silva-Ford, "Sounds of Writing and Learning."

3   See Freire, *Pedagogy of the Oppressed*. For a historical overview of the industrial model of education and the critical potentials of digital pedagogy, see Davidson, *Now You See It*.

4   Emphasis on writing, Silva-Ford notes, stems from the association of writing with linear organization of ideas and print-based media ("Sounds of Writing and Learning"). Within this design paradigm, forms of student work that are not print-based, linear expressions of arguments—including audiovisual, interactive, networked, and born-digital—are undervalued. It is worth noting that this perspective reduces the interactive potentials of textual engagements.

5   The Digital Music-Cultures course provides a critical and hands-on environment for students to explore how current music as "digital vernacular" differs from its analog, historical counterparts; how contemporary digital music-cultures create new meanings of place and identity in the increasingly globalized world; and how social, media, and technological institutions organize twenty-first-century music participation at the dispersed, grassroots level. To learn more about the course, see the course site introduction at http://cdlrsandbox.org/wordpress/digitalmusiccultures.

6   Bolter and Grusin, *Remediation*.

7   Novak, "Sublime Frequencies."

8   Drucker, "Performative Materiality"; Kirschenbam, *Mechanisms*.

9   Hsu, "Digital Ethnography"; McPherson, "Introduction: Media Studies"; Nowviskie, "Resistance in Materials."

10 On textual materials, see Moretti, *Graphs, Maps, Trees*; on sonic materials, see Clement et al., "Sounding for Meaning."

11 I have discussed the instructions and outcomes of this sounded writing assignment in a blog post. See http://cdlrsandbox.org/wordpress/racegenderpop/student-projects/musical-autobiography (accessed January 14, 2018).

12 The prompts in the assignment include: Which quotations in the Attali reading "exemplify or reflect an ideology (related to music, society, consumption, or technology) expressed by participants of the chip music community; demonstrate a technological or musical practice seen in the chip music community; differ from or challenge how music (or noise) is understood in the chip music community; explain or encapsulate the meaning of chip music in a particular social context?"

13 The instructions for this in-class activity, along with the shared document created by the students, are posted on the course site: see http://cdlrsandbox.org/wordpress/digitalmusiccultures/lessons/week-7 (accessed January 14, 2018).

14 The actual instructions used for this assignment are posted on the course

site: see http://cdlrsandbox.org/wordpress/digitalmusiccultures/assignments/assignment-7 (accessed January 14, 2018).
15 Gann and Samuels, "Deformance and Interpretation."
16 Sample, "Notes toward a Deformed Humanities."
17 Sample, "Notes toward a Deformed Humanities."
18 Attali, "Noise," 37, in Sterne, Sound Studies Reader.
19 Wayne Marshall's work claiming mashup as a pedagogical practice touches on the idea of media production as a reflexive pedagogy. Marshall's formulation seems promising because it positions digital (music) making as a critical practice, one that highlights a self-conscious engagement with the makers' personal reactions to the musical components of a mashup composition. Unfortunately Marshall's theorization falls short on its implications for classroom learning. His definition of pedagogy assumes a broad understanding of pedagogy as a transmission of knowledge between performers and audience and by extension, between scholars. See Marshall, "Mashup Poetics."

    I should note that reflexivity has surfaced in digital humanities as a part of the theorization of virtuality and human-machine interface (Hayles, *How We Became Posthuman* and *Electronic Literature*) as well as cultural rhetorics (Sano-Franchini, "Cultural Rhetorics"). A related concept of recursivity has been tied to the discourse about the public within the open-source community in anthropological literature (Kelty, *Two Bits*). These references, however, do not engage with the process of knowledge transmission in the social and sensory realms in ways that would be productive for a discussion about sound pedagogy.
20 Hahn, *Sensational Knowledge*, 10.
21 Hahn, *Sensational Knowledge*, 10.
22 Hsu, "Digital Ethnography."
23 Makagon and Neuman, *Recording Culture*, 15.
24 Ratto, "Critical Making," 253.
25 Balsamo, "Videos and Frameworks," cited in Sayers, "Tinker-Centric Pedagogy," 282; Latour, *Pandora's Hope*.
26 Low and Sonntag, "Towards a Pedagogy of Listening."
27 Stommel, "Critical Pedagogy."
28 The prompt for this workshop session is posted on the course site: see http://cdlrsandbox.org/wordpress/digitalmusiccultures/workshops/annandale-workshop-1 (accessed January 14, 2018).
29 The reflective assignment prompt is posted on the course site: see http://cdlrsandbox.org/wordpress/digitalmusiccultures/assignments/final-project-assignment (accessed January 14, 2018).
30 idanxfi, "Sounds of Learning."
31 idanxfi, "Sounds of Learning."
32 Siege, "Blooming Flowers."
33 Siege, "Blooming Flowers."

34 Siege, "Blooming Flowers."
35 Sample, "Notes toward a Deformed Humanities."
36 Siege, "Blooming Flowers."
37 Posner, Selfies, Snapchat, and Cyberbullies. Posner states the rationale for her contract grading policy on her course website. See http://miriamposner.com/dh150w15/contract-grading (accessed January 14, 2018). For more on contract grading, see Danielewicz and Elbow, "Unilateral Grading Contract."

WORKS CITED

Attali, Jacques. *Noise: The Political Economy of Music*. Minneapolis: University of Minnesota Press, 1985.

Attali, Jacques. "Noise: The Political Economy of Music." In *The Sound Studies Reader*, edited by Jonathan Sterne, 29–40. New York: Routledge, 2012.

Balsamo, Anne. "Videos and Frameworks for 'Tinkering' in a Digital Age." Spotlight on Digital Media and Learning. 2009. Accessed December 12, 2014. http://archive.is/XY3Hw.

Bolter, Jay David, and Richard Grusin. *Remediation: Understanding New Media*. Cambridge, MA: MIT Press, 1999.

Cavicchi, Daniel. *Listening and Longing: Music Lovers in the Age of Barnum*. Middletown, CT: Wesleyan University Press, 2011.

Clement, Tanya, David Tcheng, Loretta Auvil, Boris Capitanu, and Megan Monroe. "Sounding for Meaning: Using Theories of Knowledge Representation to Analyze Aural Patterns in Texts." *Digital Humanities Quarterly* 7, no. 1 (2013): n.p. www.digitalhumanities.org/dhq/vol/7/1/000146/000146.html.

Danielewicz, Jane, and Peter Elbow. "A Unilateral Grading Contract to Improve Learning and Teaching." *College Composition and Communication* 61, no. 2 (2009): 244–68.

Davidson, Cathy. *Now You See It: How Technology and Brain Science Will Transform Schools and Business for the 21st Century*. New York: Penguin, 2012.

Drucker, Johanna. "Performative Materiality and Theoretical Approaches to Interface." *Digital Humanities Quarterly* 7, no. 1 (2013): n.p. http://digitalhumanities.org/dhq/vol/7/1/000143/000143.html.

Freire, Paulo. *Pedagogy of the Oppressed*. Translated by Myra Bergman Ramos. New York: Continuum, 2005.

Gann, Jerome, and Lisa Samuels. "Deformance and Interpretation." In *Poetry and Pedagogy: The Challenge of the Contemporary*, edited by Joan Retallack and Juliana Spahr, 151–80. New York: Palgrave Macmillan, 2006.

Hahn, Tomie. *Sensational Knowledge: Embodying Culture through Japanese Dance*. Middletown, CT: Wesleyan University Press, 2007.

Hayles, N. Katherine. *Electronic Literature: New Horizons for the Literary*. Notre Dame: University of Notre Dame Press, 2008.

Hayles, N. Katherine. *How We Became Posthuman: Virtual Bodies, Cybernetics, Literature, and Informatics*. Chicago: University of Chicago Press, 1999.

Hsu, Wendy F. "Digital Ethnography toward Augmented Empiricism: A New Methodological Framework." *Journal of Digital Humanities* 3, no. 1 (Spring 2014): n.p. http://journalofdigitalhumanities.org/3-1.

Hsu, Wendy F. MUSIC112 Digital Music-Cultures. Course title. Music Department, Occidental College, Los Angeles, 2013. Accessed November 19, 2014. http://cdlrsandbox.org/wordpress/digitalmusiccultures.

idanxfi. "Sounds of Learning—Adventuring Home." MUSIC112 Digital Music-Cultures, 2013. Accessed January 5, 2015. http://cdlrsandbox.org/wordpress/digitalmusiccultures/2013/05/07/sounds-of-learning-adventuring-home.

Kelty, Chris. *Two Bits: The Cultural Significance of Free Software*. Durham, NC: Duke University Press, 2008.

Kirschenbaum, Matthew G. *Mechanisms: New Media and the Forensic Imagination*. Cambridge, MA: MIT Press, 2008.

LaBelle, Brandon. *Acoustic Territories: Sound Culture and Everyday Life*. New York: Bloomsbury, 2010.

Latour, Bruno. *Pandora's Hope: Essays on the Reality of Science Studies*. Cambridge, MA: Harvard University Press, 1999.

Low, Bronwen E., and Emmanuelle Sonntag. "Towards a Pedagogy of Listening: Teaching and Learning from Life Stories of Human Rights Violations." *Journal of Curriculum Studies* 45, no. 6 (2013): 768–89. https://doi.org/10.1080/00220272.2013.808379.

Makagon, Daniel, and Mark Neuman. *Recording Culture: Audio Documentary and the Ethnographic Experience*. Thousand Oaks, CA: Sage, 2009.

Marshall, Wayne. "Mashup Poetics as Pedagogical Practice." In *Pop-Culture Pedagogy in the Music Classroom: Teaching Tools from American Idol to YouTube*, edited by Nicole Biamonte, 307–16. Lanham, MD: Scarecrow Press, 2011.

McPherson, Tara. "Introduction: Media Studies and the Digital Humanities." *Cinema Journal* 48, no. 2 (2009): 119–23. https://doi.org/10.1353/cj.0.0077.

Moretti, Franco. *Graphs, Maps, Trees: Abstract Models for a Literary History*. London: Verso, 2007.

Novak, David. "Sublime Frequencies of New Old Media." *Public Culture* 23, no. 3 (2011): 603–34. https://doi.org/10.1215/08992363-1336435.

Nowviskie, Bethany. "Resistance in Materials." Nowviskie.org, 2013. Accessed December 31, 2014. http://nowviskie.org/2013/resistance-in-the-materials.

Owens, Paul, dir. *Blip Festival: Reformat the Planet*. Documentary. San Francisco: 2 Player Productions, 2008.

Posner, Miriam. *Selfies, Snapchat, and Cyberbullies: Coming of Age Online*. Course title. University of California, Los Angeles, 2015. Accessed January 5, 2015. http://miriamposner.com/dh150w15.

Ratto, Matt. "Critical Making: Conceptual and Material Studies in Technology and Social Life." *Information Society* 27 (2011): 252–60. https://doi.org/10.1080/01972243.2011.583819.

Sample, Mark. "Notes toward a Deformed Humanities." Samplereality.com, 2012. Accessed December 30, 2014. www.samplereality.com/2012/05/02/notes-towards-a-deformed-humanities.

Sano-Franchini, Jennifer. "Cultural Rhetorics and the Digital Humanities: Toward Cultural Reflexivity in Digital Making." In *Rhetoric and the Digital Humanities*, edited by Jim Ridolfo and William Hart-Davidson, 49–64. Chicago: University of Chicago Press, 2015.

Sayers, Jentery. "Tinker-Centric Pedagogy in Literature and Language Classrooms." In *Collaborative Approaches to the Digital in English Studies*, edited by Laura McGrath. Logan: Utah State University Press, 2011. Accessed December 12, 2014. http://ccdigitalpress.org/ebooks-and-projects/cad.

Siege. "Blooming Flowers: A Remixed Journey of Vocal Discovery." MUSIC112 Digital Music-Cultures, 2013. Accessed January 5, 2015. http://cdlrsandbox.org/wordpress/digitalmusiccultures/2013/05/06/blooming-flower-a-remixed-journey-of-vocal-discovery.

Silva-Ford, Liana. "The Sounds of Writing and Learning." *Sounding Out!*, August 27, 2012. http://soundstudiesblog.com/2012/08/27/the-sounds-of-writing-and-learning.

Stommel, Jesse. "Critical Pedagogy: A Definition." *Hybrid Pedagogy*, 2014. Accessed November 19, 2014. www.hybridpedagogy.com/journal/critical-digital-pedagogy-definition.

Winter, Robert, et al. History of Civilizations. Course title. Occidental College, Los Angeles, 1969. Special Collections. Accessed on March 21, 2013.

III

DISCIPLINARY TRANSLATIONS

07

WORD. SPOKEN.

Articulating the Voice for High Performance Sound Technologies for Access and Scholarship (HiPSTAS)

TANYA E. CLEMENT

Now accessing audio online seems easy. We find what we want to listen to through Google or through a search box on a favorite site. We can click on a link, open the file right in the browser, and then press play, fast-forward, and playback. In some cases, we can even view the sound waves or spectrograms associated with the audio or we can annotate what we hear and remix these representations. At the same time, modes of computational analysis with sound that let us search for sounds with sound or map sonic patterns across collections of audio, for example, remain few and relatively simplistic.

 The editors of this collection have rightly asserted that digital sound studies must include technology as an object of study in order to attend to "the ways that various devices mediate sound, from the speaker and microphones to software coding and hardware development" (introduction). Using technologies to enhance access to and analysis of audio collections seems to promise a wide range of critical "close" and "distant" critical listening opportunities in digital sound studies, but there are still few conversations about the many ways in which digital infrastructure technologies, or the hardware and software that facilitate these methods, influence scholarship.[1] To better understand these mediations in the context of developing tools for critical listening, this chapter considers classification systems for sound as a significant object of study for better understanding the digital infrastructure technologies that facilitate scholarship with audio.

Technologies used to facilitate scholarship with audio require a classification system to "mark" or annotate features of digital audio or text so that we can organize and search them more easily. By limiting the computer's search to identifying keywords or concepts such as an author name, a date range, or a genre (like horror or comedy, for example), we get expected results more quickly. Though they often seem invisible in the digital realm, classification systems reflect how we interact with machines as social and situated beings. Classification systems are subjective and deeply political: one person's horror movie could be another's comedy.

Classification or standardization protocols are subjective and political because they are sociotechnical phenomena—pertaining to both human and technical influences. A sociotechnical perspective sees technologies as "ways of life, social orders, practices of visualization" that are interdependent with the politics of knowledge production.[2] From this perspective comes the understanding that the classification standards we develop, which ultimately shape the knowledge produced through them and by them, are developed according to our own perceptions of the world.[3] So, while we need standardized protocols such as classification systems to make our hardware and software work more efficiently for everyone, we also need to learn how to interrogate these systems in order to understand how our assumptions and biases impact the knowledge we produce with these technologies.

To frame this study from a sociotechnical perspective and within the particularities of digital sound studies, this chapter considers a specific digital humanities project in sound—High Performance Sound Technologies for Access and Scholarship (HiPSTAS)—and a particular aspect of development within that project—the use of standardized classifications for describing sound features within the development of a tool for searching sound with sound.[4] Situating this aspect of development within HiPSTAS within a brief history of methods for classifying sound features will help us consider the impact that technology and politics can have in shaping scholarship in digital sound studies.

Sound in the HiPSTAS Project

A joint project of the School of Information at the University of Texas at Austin and the Illinois Informatics Institute at the University of Illinois at Urbana-Champaign, HiPSTAS was initially funded by the National Endowment for the Humanities as an Institute in Advanced Technologies in the Digital Humanities.[5] The HiPSTAS Institute included twenty junior and senior faculty and advanced graduate students as well as librarians and archivists in the humanities from across the U.S. interested in analyzing large collections of spoken-word audio collections using high-performance or "supercomputing" technologies. Among many collections of interest to the participants were 30,000 files of recordings from PennSound's poetry archive; 600,000 digital collections objects from the American Folklife Center at the Library of Congress; 30,000 hours of oral histories from StoryCorps; and 3,000 hours in the American Philosophical Society's Native American Collection, which includes recordings from more than fifty tribes across North America, among other collections. The participants met in two face-to-face meetings in May 2013 and May 2014 as well as in monthly virtual meetings. The objectives of the HiPSTAS Institute were threefold: first, to assess how these communities wanted to use computational tools to study spoken-word collections; second, to assess how those tools needed to be developed to support analyzing and visualizing large audio collections in the humanities; and third, to produce preliminary results with these tools using the collections of interest to the participants.

A significant aspect of the HiPSTAS Institute included introducing the participants to the Adaptive Recognition with Layered Optimization (ARLO) software. ARLO, which was originally developed by HiPSTAS co-PI David Tcheng for acoustic studies in animal behavior and ecology, had previously been used to search for bird calls across field recordings. Conceived to model a bank of hairs in the inner ear, which vibrate at different audio frequencies in response to sound waves, ARLO monitors and then samples each "hair's" instantaneous energy (a sum of the tuning fork's potential energy or the deflection of the fork and its kinetic energy based on the speed of the movement, per second). ARLO uses this data to create a 2D matrix of values (frequency vs. time) called a spectrogram. Essentially, these spectrograms (see fig. 7.1) show a map of sonic energy across time: each row of pixels represents a frequency band, and the color of each pixel represents the numeric value of total energy of that particular frequency (or how much the tuning fork trembles) for that point in time. ARLO uses these spectrograms to extract sonic features for machine-learning processes, including unsupervised learning such as clustering as well as supervised learning for classification.[6]

Used to search across sound collections for sonic patterns, these machine-learning processes rely on human intervention. To teach the software to identify sounds of interest with supervised learning techniques, human "experts" annotate the sounds they want to find and use these seed examples to teach an algorithm to find other, similar sounds. With unsupervised techniques, the "expert" still chooses certain features of the audio to guide how the machine-generated clusters are formed. Thus, software like ARLO finds sounds by comparing each training example to new, unlabeled examples and determining good matches as those that seem to have some of the same features, such as the total energy value described above. For the ornithologist who is examining thousands of hours of birdcalls, this process of matching might mean marking (or "tagging") examples of a particular bird's call on a spectrogram and asking the software to retrieve similar calls. In the case of a humanist, such as one of the scholars at the HiPSTAS Institute, this could mean tagging moments of laughter, applause, gunshots, or feedback noise to teach the machine to find more such events. In each case, the machine is taught with these seed examples to find or cluster what the expert has marked as interesting.

Machine-learning software like ARLO relies on many seed examples to train the algorithm. Consequently, realizing that the participants could produce more and possibly better seed examples if they worked together or with students, we developed a collaborative interface for tagging example sounds. Figure 7.1 shows the tagging interface we created for participants interested in analyzing the PennSound poetry archive.[7] The interface provides the listener with a two-second sample that has been randomly selected from PennSound's approximately 5,500 hours of audio. The listener chooses labels to apply to the sample and then receives the next example. In this way, the listener can easily and quickly "mark up" a collection with examples for machine learning.

The most significant aspect of this example for this discussion concerns how we chose the labels we used in the tagging interface. The tagging interface reflects a classification schema or set of rules that the PennSound poets and scholars chose for labeling the sound snippets.[8] They chose the classification schema, found in the "Transcriptions of Speech" section of the Text Encoding Initiative (TEI) P5 Guidelines for Electronic Text Encoding and Interchange, for conceptual and practical reasons. First, they chose this schema because they wanted classifications that reflected the patterns they

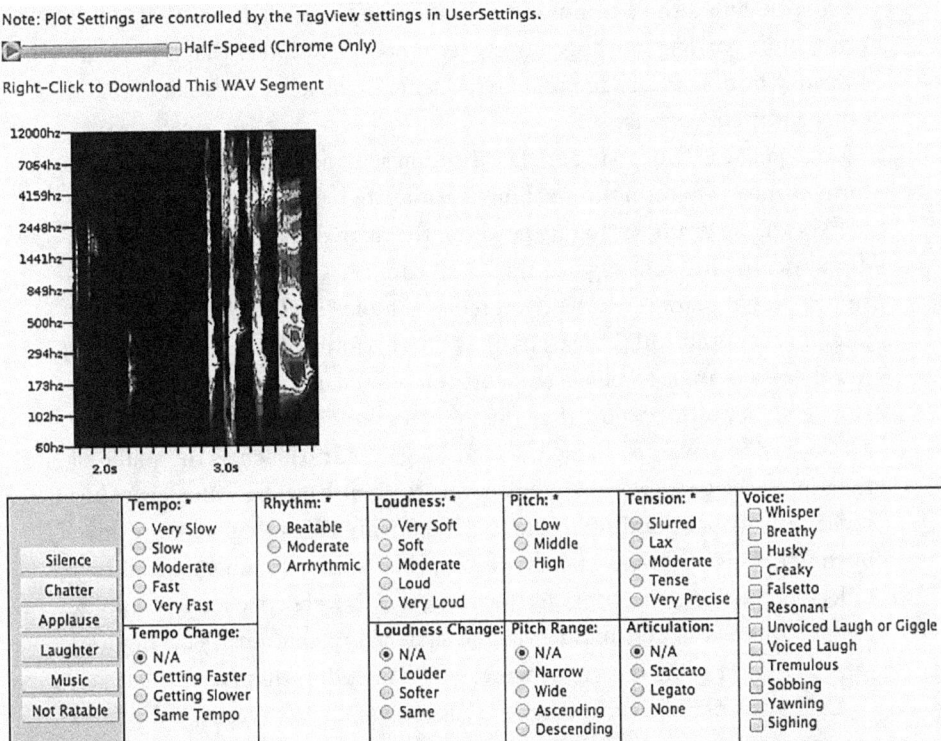

FIGURE 7.1 A tagging interface used to classify sound features on examples from PennSound.

sought to discover in their collection. In particular, the poets and scholars analyzing the PennSound collection were interested in analyzing the "vocal gestures" that Charles Bernstein (PennSound codirector) has argued "are available on tape but not page" and "are of special significance for poetry": namely clusters "of rhythm and tempo (including word duration)" and "of pitch and intonation (including amplitude), timbre, and accent."[9] By using descriptors from the TEI Transcription for Speech guidelines, the PennSound participants believed they had found terms that accurately described what they were hearing and what they wanted to find.

Second, the PennSound participants wanted a classification schema or standard that had been vetted by peers and that held the promise of facilitating future collaborations among projects that had already used (or might in the future use) these classifications. Released in November 2007, TEI P5 is a broad set of guidelines for an XML schema that is in wide circulation in the digital humanities community. By using the TEI labels or schema, the PennSound participants hoped to create a set of descriptors that they might someday be able to use to compare classifications across PennSound and other audio collections.

The goal was to use the TEI classification schema to facilitate many uniform examples to train the machine-learning algorithm. Given a collection of two-second examples to tag, however, the thirteen participants assigned dramatically different tags to the same sample. One participant, for instance, might mark the same two-second sample "Beatable" with a "High" pitch and another might classify it as "Arrhythmic" with a "Low" pitch. Another issue arose when participants wanted to label contexts rather than snippets; they wanted more than the two-second window they were given by ARLO, and they wanted to tag the recording scenario (such as the sound of the room), the gender of the speaker, and the genre (such as music) as they perceived it, not according to the specified genre types that were provided by the TEI classification schema. That is, they wanted to label *the label* as it reflected their own listening perspectives, which were couched in complex understandings of culture, genre, and materiality, but our ARLO tagging interface, built using the TEI schema, would not allow them to do that.

Using this defined vocabulary or schema, which was meant to facilitate the process by providing uniformity across the examples, the PennSound participants debated how and when to implement the classifications. The PennSound participants struggled with labeling what they saw on the spectrograms, often citing doubts about their ratings and their understandings of the classifications and especially showing a resistance to the TEI classification system they had chosen to use. While classifying snippets of sound seemed to work well for the ornithologist, classifying snippets of poetry performances according to the chosen standard seemed to frustrate the humanist's desire to find dynamic or time-based aspects of performance. It seemed that while the sound of a bird could more easily be classified as "male cardinal," classifying or defining the human voice—an act that Jonathon Sterne calls a debate over "what it means to be human"—was a more provocative endeavor.[10] Realizing on the one hand that the classification system was necessary for increased computational productivity and efficiency but also, on the other hand, that it was flawed in its orientation, the PennSound participants did not seek to discard the use of a classification system but rather cited the need for a "better" (i.e., more accurate) classification system for describing sonic features as a high-priority requirement for moving ahead with developing ARLO.

I tell this story because it provokes sociotechnical questions for digital sound studies in general. How can a classification system, which is an infrastructural mainstay for facilitating computational analysis, mediate knowledge production? And how can we study these mediations? Bowker and Star suggest "infrastructural inversions" as a method for better understanding these interdependences between standardizations and knowledge production.[11] As the authors suggest, we must take into account that standardized classifications and systems are ubiquitous; they are both materially and symbolically realized as well as historically situated, representing multiple voices and silences.[12] Ultimately, classification systems reflect philosophies concerning the nature of sound as well as the practical politics involved in developing such standards that include what remains visible and invisible in the system.[13] In my example above, we see an example of how a classification system might work in a tool like ARLO. The next two sections consider the historic roots of this system to better understand why they might have seemed inaccurate or inappropriate to the PennSound scholars. In particular, I will consider the symbolic and material underpinnings of the TEI's Transcriptions of Speech classifications for sound within the history of philosophies in linguistics and the immediate political contexts that affected the establishment of these standardized rules.

A Brief Look at Prosodic and Paralinguistic Classifications

Linguists have been at the forefront of establishing complex and standardized protocols for describing spoken language. Driven by the desire to address the "practical needs of spoken language corpora annotation and analysis," especially in the light of more recent developments in computer-facilitated speech analysis, Maciej Karpiński outlines seventy-five years of research in linguistics concerning attempts to define and categorize what we say and how we say it.[14] In a specific example that is of particular use for this discussion, linguists often use "prosody" as a phenomenon comprising varying degrees of intonation, stress, and rhythm that convey meaning through phrasing and prominence, while they describe paralinguistic features as those that do not easily belong to a describable linguistic structure.[15] David Crystal and Randolph Quirk divide their seminal study *Systems of Prosodic and Paralinguistic Features in English* (1964) into prosodic and paralinguistic features based on how easily these features might be integrated into typical linguistic structures.

Specifically, Karpiński claims that prosody may be measured or described using three basic parameters—pitch frequency, duration, and intensity—and that these parameters influence each other as communicating features.[16] In written texts, prosodic features are typically described in terms of syntactical units. These language features often include parts of speech, accent, phoneme, stress, and tone as well as other information that influences how a sentence can be read such as the position of a word in a phrase (e.g., consecutive verbs or multiple nouns), sentence type (e.g., a declaration or a question), and information structure (e.g., independent versus dependent clauses, since inferable information in a dependent clause is usually deaccented).[17] In other words, when we seek to "sound out" a written word, we guess how to pronounce words unknown to us based on our experiences with prosodic features such as recognizable clues for pronunciation in the surrounding syntax. Nouns in a series require different amounts of stress, for instance, and questions have a lilt.

In comparison, paralinguistic features seem more difficult to describe and standardize. In his attempt to delineate terms, for example, Karpiński discusses paralinguistics within the context of three areas of study that include prosody, vocal quality, and gesture.[18] Also referred to as *timbre*, voice quality in musical instruments connotes the distinctive sound a particular instrument makes in contrast to another—such as the sound of an oboe versus that of a tuba—even when the instruments are playing the same note at a similar amplitude. For Karpiński, such vocal features are "individual, idiosyncratic, and further from 'language proper'" than prosodic features, making them "multidimensional and difficult to operationalize."[19] Crystal and Quirk also note the difficult and slippery nature of categorizing paralinguistic vocal qualities that surround such sounds as giggling, laughing, and crying:

> It is not possible to say when *giggle* ends and *laugh* begins, or when *cry* ends and *sob* begins, though doubtless it would be possible to examine a great quantity of data and obtain some measurements (of pulse speed, air pressure, prominence, for example) which would be of value in establishing more objective gradations.[20]

It is useful to note that Crystal and Quirk, who have attempted to systematize these voice-quality measures in *Systems*, put prosodic features on the more "describable" end of the classification continuum from prosodic to paralinguistic, even while they are quick to note that there is no sharp division between them. "It is doubtful," they write about implementing a system of vocal-quality categories, "whether the results would justify the time and ingenuity involved."[21]

Certainly, how we perceive and make meaning with prosodic and paralinguistic features is a subjective activity. Dwight Bolinger asserts that intonation "is generally used to refer to the overall landscape, the wider ups and downs that show greater or lesser degrees of excitement, boredom, curiosity, positiveness, etc."[22] Further, in its expansiveness, prosody can signify elements of a speaker's identity including affect and emotional engagement, age, cognitive process and development, ethnicity, gender, and region and has been used to study human behavior, culture, and society.[23] For these reasons, Karpiński points out, prosodic and paralinguistic features are often considered "indexicals" since they seem to point to the context of a person or place.[24] Indeed, Karpiński describes paralinguistics such as laughter, giggles, gasps, pauses, hesitations, or coughs as "all the phenomena and features of a speaker's behaviour that go beyond the (current) limits of systematic linguistic description but still influence the way his/her communicational contribution is understood by his/her conversational partner."[25]

Tasked with submitting recommendations for the TEI's Transcriptions of Speech section of the guidelines, then, the TEI Spoken Text Working Group (STWG) relied on Crystal and Quirk's *Systems* and its assertions that prosodic and paralinguistic features influence meaning-making with sound as a basis for identifying which speech characteristics in recordings should be (and could be) marked in the guidelines.[26] This Crystal and Quirk perspective is reflected in TEI labels that include the following attributes:

- *Tempo:* Very Slow, Slow, Moderate, Fast, Very Fast
- *Rhythm:* Beatable (highly rhythmic), Moderate, and Arrhythmic (flat or ordinary speech)
- *Loudness:* Very Soft, Soft, Moderate, Loud, and Very Loud
- *Pitch:* Low, Middle, and High
- *Tension:* Slurred or Lax (for looser articulation), Very Precise or Tense (for pronounced articulation)
- Other classifications include Whisper, Breathy, Husky, Creaky, Falsetto, Resonant, Unvoiced Laugh or Giggle, Voiced Laugh, Tremulous, Sobbing, Yawning, and Sighing.

Notably, this list is not an exact reflection of Crystal and Quirk's work, in which voice *qualities*, which include different modes (normal voice, whisper, breathiness, huskiness, creak, falsetto, and resonance); and voice *qualifications*, which ordinarily interrupt speech (laughter, giggling, tremulousness, sobbing, and crying) are considered separately.[27] In contrast, the TEI guidelines foreground the similarities between voice qualities and qualifications by grouping them together in a list of "other" classifications.

It is these voice-quality features, which are regarded by linguists as difficult to systematically categorize, that the HiPSTAS project participants found most compelling in their attempt to systematically annotate their spoken-word recordings. The voice quality or timbre aspects of paralinguistics, which Bernstein calls the "poet's aesthetic signature or acoustic mark," are particularly important in studying poetry performances.[28] As an indexical property, they appear in a spoken poem or performance as "a technical feature that can be used to form or deform social distinctions and variations."[29] Consequently, as mentioned, the PennSound scholars chose to adopt the TEI descriptors for philosophical reasons, because the terms, adopted from Crystal and Quirk, seemed to reflect their own concerns, but they also chose them for practical reasons, since they had been adopted by an authority (the TEI community) and seemed to promise some consistency across projects, authors, and poems of interest as well as offering future possibilities for collaboration with other projects using the TEI guidelines. The advantages that come with building such a system, however, belie not only practical concerns about the fact that marking up audio takes time and resources but also philosophical concerns as to the erasure of a long history of conversations about the subtle differences between voice quality and qualifications.

Three Compromises for Classifying Sound

Bowker and Star suggest a means by which we can better articulate the sociotechnical nature of classification systems. Defining such systems as "a rich set of negotiated compromises ranging from epistemology to data entry that are both available and transparent to communities of users," they challenge scholars to make such compromises readily apparent for consideration.[30] A primary compromise of interest for digital sound studies is one the introduction to the TEI guidelines articulates well: "An electronic representation must strike a balance between the following two, partially conflicting, requirements: authenticity and computational tractability."[31] *Authenticity*, in this sense, is subjective and corresponds to whether or not a digital surrogate or representation seems "true" or accurate to a philosophy about or understanding of that phenomenon in the world. *Computational tractability* is the extent to which that representation is computable or representable in the computational environment, which includes the software, the platform, the hardware, and the networks being used to consider that representation. Thus, a philosophical concern for what is authentic in a community of scholars such as digital sound studies scholars must be in constant conversation with practical concerns for what is computationally tractable in a digital environment.

By learning to articulate the nature of these sometimes conflicting requirements (at once philosophical and practical), we are empowered in the digital sound studies community to impact how the systems used by the community are designed and implemented. Below, based on a close look at the history of how the paralinguistic voice qualities in TEI's Transcriptions of Speech schema came to be, and a consideration of how the HiPSTAS participants attempted to apply these guidelines with the ARLO software, I have suggested three more general areas of compromise for consideration in digital sound studies.

COMPROMISE #1: Moving from Text to Sound

The first compromise for consideration is one that balances a desire for "user friendly apps" against a desire for applications or software that fully represent the subtle characteristics of a phenomenon. We are used to polished and seemingly intuitive applications for searching, browsing, publishing, and teaching with text, but applications for searching, browsing, publishing, and teaching with sound are emergent, developing, and often "buggy." In such a context, we must consider the compromises inherent in choosing ease-of-use technologies over change-of-paradigm technologies.

For example, when the TEI STWG was tasked with submitting recommendations for the TEI's Transcriptions of Speech section of the guidelines, they focused on guidelines for marking up text-based transcriptions of recordings rather than guidelines for the faithful representation of the recordings' many sonic attributes.[32] This focus was the result of STWG's perspective on prosodic and paralinguistic features as problematic, such as "speaker overlap, pauses, hesitations, repetitions, interruptions," uncertainty, and context.[33] It is clear from citations in their extant working notes and drafts that the STWG were versed in the works of Svartvik and Quirk ("A Corpus of English Conversation") and Tedlock (*The Spoken Word*) and considered paralinguistic and prosodic sound features expressive; yet based on the need to make a hierachical representation of text a main tenet of TEI, they found these time-based and overlapping sound dynamics—such as pitch, speed, and tone—impractical to represent. (Indeed, encoding for recorded speech was not included at all in the original TEI P1 guidelines.)[34] In short, the STWG's theoretical or philosophical understanding of sound did not coordinate well with the means they had to express or represent this understanding. Notes from the working group's 1991 meeting reflect the compromises they knew they were making:

> In a brief discussion on performative features such as pitch, speed and vocalisation, LB [Lou Burnard] asked if these could not be regarded as analogous to rendition in written texts and treated in a similar way. It was generally felt that it would be better to mark these using milestone tags such as <tag>pitch.change</tag>, <tag>speed.change</tag> etc.[35]

The STWG concluded that topics including "quasi vocal things such as laughter, quasi lexical things such as 'mm,' prosody, parallel and discontinuous segments, uncertainty of transcription, uncertainty in general" needed "considerable further work."[36] And, these "quasi lexical things" remain peripheral to the guidelines even today.

This peripheral status is reflected materially in how these paralinguistic features are included in the TEI standards. The STWG relegated voice quality—paralinguistic characteristics such as pitch and speed, etc.—to a "shift" tag or element.[37] The "shift" element (<shift/>) is represented "as pairs of milestone tags marking positions of prominence . . . with the 'end' tag of the pair being replaced by a shift to normal."[38] The choice to use this kind of element is significant, because <shift/> requires the encoder to mark dynamic sound attributes in the encoded transcript as shifts to and from a "normal" speaking mode (see fig. 7.2).

Beyond assumptions about normativity that exist behind establishing a "normal" speaking mode, elements like the <shift/> are conceived as phenomena that happen in discrete moments of time. The <shift/> element occurs in one spot and marks specific points in a transcript, as if the dynamism of such sonic features could be pinpointed in time. Though the TEI guidelines are clear in the assertion that they are "not intended to support unmodified every variety of research undertaken upon spoken material now or in the future," the STWG's choice to show paralinguistic entities as "shifts" from a "normal" state and as discrete, "well-defined units" flies in the face of discussions by linguists such as Crystal, Quirk, and Karpiński,

```
<u>
  <shift feature="loud" new="f"/>Elizabeth
</u>
<u>Yes</u>
<u>
  <shift feature="loud" new="normal"/>Come and try this <pause/>
  <shift feature="loud" new="ff"/>come on
</u>
```

FIGURE 7.2 An example of the "shift" element from the TEI P5 guidelines.

who discuss the dynamic, subjective, and slippery nature of paralinguistic features. Indeed, the guidelines for these features were intended primarily for enabling the linguistic study of spoken text recordings as "a written or electronic representation of a stretch of speech which is treated for some purpose as a well-defined unit."[39] The material instantiation of these features in the <shift/> element shows how sound attributes become marginalized in computational infrastructures that focus on text and spoken language.[40]

Other, more recent projects for developing classification schemas for sound can help us imagine other compromises we must make in our attempts to balance our desire for the niceties of systems built for textual searches and our desire for new systems that better facilitate sonic searching. For example, the Federal Agencies Digitization Guidelines Initiative (FADGI) formed as a group in 2007 "to define common guidelines, methods, and practices to digitize historical content in a sustainable manner."[41] FADGI's metadata standard, "Embedded Metadata in Broadcast WAVE Files, Version 2," describes sonic information that points to sound's materiality, including signal chain specifics, sample rates, and bit depth. Further, other classification schemas proposed by the International Association of Sound and Audiovisual Archives (IASA) capture information concerning an audio file's provenance and historical context such as the date and place of a recording.[42] Even with these advancements, questions remain concerning the extent to which classifications such as FADGI's help us better understand vocal gestures and whether narrative descriptions of soundscapes give us enough information about sonic histories. Ultimately, these standards are works in progress and the sociotechnical histories behind the development of these standards also reflect compromises, both philosophical and practical, that organizations other than the TEI will make based on a desire to balance their situated understanding of sound, the perceived needs of the communities they serve, and the technologies they hope to employ in the service of these goals.

COMPROMISE #2: Moving from Fixed to Emergent Meanings

Another compromise to consider in digital sound studies is one that weighs a desire to represent sounds as fixed in meaning against the difficult work of representing sounds as phenomena with emergent and multiple meanings. This is a significant compromise to address because any digital representation of the experience of sound will need be, by nature, a reduction that nonetheless invites expansive thinking.

Sound studies scholars in the humanities have been primarily interested in articulating sound culture in all of its complexity rather than in simplified, linear, or atomistic terms. For instance, citing Jacques Derrida, Dennis Tedlock dismisses "the entire science of linguistics, and in turn the mythologics (or large-scale structuralism) that has been built upon linguistics," since such sciences and mythologics are "founded not upon a multidimensional apprehension of the multidimensional voice, but upon the unilinear writing of the smallest-scale articulations within the voice."[43] Michael Chion argues that a recorded artifact has fixity that is necessary for close listening since to perceive sonic traits, one must listen repeatedly to a recorded moment, but he dismisses the state of fixedness that a framework like a classification system would engage since within it sounds "acquire the status of veritable objects" and "physical data"; this fixed data, he asserts, is inauthentic since it does not represent what was actually heard within the real time of "presence."[44] Likewise, Bernstein notes that "systems of prosodic analysis" that regularize sound "break down before the sonic profession of reading: it's as if 'chaotic' sound patterns are being measured by grid-oriented coordinates whose reliance on context-independent rations is inadequate."[45] These statements reflect an understanding of sound in the humanities as an emergent phenomenon that is dynamic and in flux and that evolves and expands over time, constantly introducing ambiguity and uncertainty. As such, there is a clear resistance toward "fixing" sounds for better understanding of meaning-making processes.[46]

Reduction as a means of representation is unavoidable in a digital context, but there are choices that dictate the terms of these reductions. Most of the categorizations outlined above, for example, have been established from the perspective of linguistic study and, for the most part, in terms of creating transcriptions from audio files. In contrast, classification systems devised by poets may be designed to represent the differences between breathy or harsh voices; a system designed by historians may better show the sounds of a city venue; and one designed by Native Americans might facilitate comparing the changing paces of elders' stories. In each of these scenarios, one can imagine that certain sonic attributes are foregrounded based on the interests of a particular community.

A compromise that helps us better engage the emergent nature of sound hermeneutics in digital space is a choice that may be quite productive in digital sound studies. For instance, Kenneth Sherwood cites fixed and discrete instances of repetition as perceptible signs of emergence that signify elaboration and versioning. Bernstein notes the signification of dynamic "performative gestures" such as emotional intensification, which can map to measurable changes in heightened and decreased sound frequencies and speed.[47] Likewise, Crystal and Quirk have identified measurements for establishing the emergent dynamics of voice qualifications as "objective gradations" by "setting up parameters for degrees of pulsation types, pulsation speed, oral aspiration, nasal friction, air pressure, amplitudes and frequency of vocal cord vibration, and volume and tension of supraglottal cavities."[48] These examples demonstrate that the dynamics of a voice—its increasing or decreasing pace, its tone changes over time—can be understood against different frames of reference that we may choose to position as "fixed" (such as the words of a poem) even as we understand them to be in flux. This choice against fixity can be forwarded by classifications that help us better articulate and understand the terms of fixity as choices.[49] As such, sound as a phenomenon of emergence could be understood in terms of how it is represented as fixed.

COMPROMISE #3: Moving from Discrete to Contextual

A third compromise for consideration in digital sound studies entails balancing the desire for representing sound as a discrete event in time with the difficult work needed for describing sound across particular time contexts. In the previous section, I argue that we must represent fixed points in the sonic event in order to study the emergences of meaning. We must also better understand how we represent these features in fixed moments of time. Repeated, elaborated, or intensified moments can be marked as discrete, for example, even as their significance is based on their relationships across time with other moments. By constellating fixed moments in relationship to each other and situating them as patterned contexts over time, then, we may do the difficult work that we must do to develop classification systems that use fixity as means for representing contextualization and emergence.

Current guidelines for describing the historical context of recordings can provide an example of such a compromise between fixed and fluid representations. The TEI, for instance, includes a "recording" element that allows the encoder to describe dates; times of day; statements of responsibility for authors, editors, producers, etc.; the recording equipment used; or whether a broadcast recording is the basis of the text being transcribed and described. As well, there is a provision for adding elements that also describe the "setting" of a recording and its "participants." The FADGI guidelines contain these fields as well as the "bext chunk," which holds data on the digitizing process (including the analog source recording), on the capture process, on information about the storage of the file, and on versions of the coding history related to the file itself.[50] In many cases, these are optional fields that remain empty even as this contextual information impacts how we perceive the relationships that are marked and ultimately what and how we hear.

The choice to include this kind of contextual information reflects a desire to articulate design standards that do not just report on relationality but rather encode it. Innovative and productive work for representing relationality is already happening in the context of speech transcriptions. IASA recommends the Resource Description Framework (RDF), a World Wide Web Consortium (W3C) specification that allows humanists to describe relationships between objects on the web. Currently used by ARC (Advanced Research Consortium), for example, to provide gateways and venues for peer-reviewing digital projects for a variety of disciplines in literary study, RDF facilitates searching and finding relationships across projects in Networked Infrastructure for Nineteenth-Century Electronic Scholarship (NINES), 18thConnect (focused on eighteenth-century scholarship), the Medieval Electronic Scholarly Alliance (MESA), Renaissance Knowledge Network (ReKN), and Modernist Networks (ModNets). Using RDF, the ARC infrastructure is powerful, because each of the ARC nodes has its own stand-alone interface, but all of the resources can be searched together through the ARC catalog. A search on NINES can be modified to find objects from MESA, for example. While further work needs to be done to imagine an RDF schema that reflects relationships across sonic features of interest, using something like the ARC infrastructure for sound files could mean cross-searching that includes sound resources, too, which in turn would enable scholars to better collaborate on digital audio projects on a local and global scale across disciplines and interests. This kind of relationality is similarly the future of new International Image Interoperability Framework

(IIIF) guidelines for facilitating better access to audio collections through application programming interfaces (APIS).

A clear next step is to build tools that facilitate the ability to act on encoded relationality. Karpiński, for instance, proposes a "coherent approach" to linguistic data annotation that would take into account the indiscrete nature of speech prosody and voice-quality features.[51] Recommending that we treat these features as continua rather than categories, Karpiński argues that prosodic and paralinguistic features are multifunctional, multimodal, and multileveled, as well as both global and local; as such, he recommends implementing "sliders or joysticks for data input and to refrain from imposing any points on the scale, such as from a stable to a trembling voice, with all intermediate states possible."[52] Thomas Schmidt also suggests practical solutions for implementing varied perspectives on sound, such as bringing seven tools that are most commonly used by linguists for spoken language transcriptions—ANVIL, CLAN/CHAT, ELAN, EXMARaLDA Partitur-Editor, FOLKER, Praat, and Transcriber—to a common TEI schema.[53] Karpiński's and Schmidt's interventions suggest compromises that encompass the practical issues related to a need for discrete categorizations, such as annotations for linguistic data or a TEI schema, with the need to represent relationships across perspectives from multiple communities in multiple contexts across time.

Conclusion

Classification schemas for sound are language-based: they are themselves texts that attempt—sometimes with frugal and other times with rich results—to describe the world of sound that is always beyond text, beyond a listener, beyond one single snippet of a recording played back at one point in time. To approach the complexities that characterize our experiences with sound, there are many more philosophical and practical compromises we will have to negotiate as we continue to develop productive infrastructures for digital sound studies. We will need to consider what it means to engage sound thoughtfully, expansively, and critically with computational instruments that are often modeled on the normative practices of "hearing" with the ear when the ear is not the only hearing instrument. "I can hear more plainly through my teeth than through the external ear," Thomas Edison admits; "A stick touching a music box and placed between my teeth enables me to enjoy the music."[54] Another compromise will entail balancing well-intentioned plans to incorporate crowd-sourced listener responses with the practical need for clean and manageable digital sound data. Tsur reminds us, for instance, that "sophisticated electronic instruments do give an accurate analysis of the sound information; but what really matters is its integration as it takes place in the brain" of each listener.[55] We must learn to balance this desired sophistication with the vast amount of data that a systems manager or a researcher would then have to manage, process, clean, and analyze. The technologies we are using are situated, personal, and political, but they also require practical interventions for use; they are indeed ways of life.

The ultimate compromise digital sound studies will face in negotiating authenticity and computational tractability is not new to sound studies: it includes any attempt to perceive the world outside the biases we bring to everything we do. In his 1889 article "On Alternating Sounds," for instance, Franz Boas considers the extent to which a philologist's field notes reflect the phonetics of his own language and writes that in the field, philologists "reduce to writing a language which they hear for the first time and of the structure of which they have no knowledge whatsoever. . . . Each apperceives the unknown sounds by the means of the sounds of his own language."[56] Indeed, it is the compromises that we will make as we model, engage, and interpret digital sound in new ways that will provide opportunities for provocation and for questioning our unavoidable biases as listeners.

NOTES

1. Bernstein, *Close Listening*; Clement, "Distant Listening."
2. Quote from Haraway, "Situated Knowledges," 583. See also Bowker and Star, *Sorting Things Out*; Bardzell and Bardzell, "Towards a Feminist HCI Methodology"; Berg, "Politics of Technology"; Feinberg, "Two Kinds of Evidence"; Frohmann, *Deflating Information*.
3. Bowker and Star, *Sorting Things Out*, 34.
4. Tanya E. Clement is the primary investigator of the HiPSTAS project. Please see more information at www.hipstas.org.
5. The project continues with a second NEH grant from Preservation and Access for Research and Development, titled HiPSTAS for Research and Development with Repositories (HRDR). Please visit www.hipstas.org.
6. Downie et al., "Novel Interface Services."
7. Launched January 1, 2005, PennSound is the largest collection of poetry sound files available for noncommercial distribution on the Internet. PennSound is

codirected by Charles Bernstein and Al Filreis and associated with the Center for Programs in Contemporary Writing and School of Arts and Sciences Computing at the University of Pennsylvania.

8 The meeting took place at the PennSound offices in Philadelphia, October 26, 2013, including the PennSound directors Al Filreis and Charles Bernstein as well as a host of their senior editors and technical advisors (Michael Henessey, Chris Martin, Steve McLaughlin, Danny Snelson, and others).
9 Bernstein, Attack of the Difficult Poems, 126.
10 Sterne, "Sonic Imaginations," 11.
11 Bowker and Star, Sorting Things Out, 37.
12 Bowker and Star, Sorting Things Out, 37–40.
13 Bowker and Star, Sorting Things Out, 44.
14 Karpiński, "Boundaries of Language."
15 Rooth and Wagner, "Harvesting Speech Datasets"; Crystal and Quirk, Systems.
16 Karpiński, "Boundaries of Language," 41.
17 Becker et al., "Rule-Based Prosody."
18 Karpiński, "Boundaries of Language." While gestures are very important to any oral performance such as poetry, they are not the focus of this study.
19 Karpiński, "Boundaries of Language," 43.
20 Crystal and Quirk, Systems, 42.
21 Crystal and Quirk, Systems, 42.
22 Bolinger, Intonation and Its Parts, 11.
23 Rooth and Wagner, "Harvesting Speech Datasets."
24 Karpiński, "Boundaries of Language."
25 Karpiński, "Boundaries of Language," 47.
26 Karpiński, "Boundaries of Language"; Schmidt, "TEI-Based Approach." TEI, TEI P5 cites Boase (London-Lund Corpus) as a reference for these materials, but this text was never published (according to Edwards and Lampert, Talking Data) and is no longer available. A conversation with Lou Burnard led me to Crystal and Quirk (Systems) as an alternative reference for these descriptions.
27 Crystal and Quirk, Systems.
28 Bernstein, Attack of the Difficult Poems, 127.
29 Bernstein, Attack of the Difficult Poems, 127.
30 Bowker and Star, Sorting Things Out, 34.
31 TEI Consortium, TEI P5.
32 See Johansson et al., "TEI AI2 M1"; "TEI AI2 M2"; and "TEI AI2 W1."
33 Johansson, "Encoding of Spoken Texts," 150.
34 Sperberg-McQueen and Bumarde, "Guidelines."
35 Johansson et al., "TEI AI2 M1."
36 Johansson et al., "TEI AI2 M1."
37 TEI Consortium, TEI P5.
38 Johansson et al., "TEI AI2 M2."
39 Johansson, "Encoding of Spoken Texts," 149. The working group's final recom-

mendations reiterate this definition: "The goal of an electronic representation is to provide *a text* which can be manipulated by computer to study the particular features which the researcher wants to focus on" (Johansson et al., "TEI A12 W1"); and the current guidelines state conclusively: "Speech regarded as a purely acoustic phenomenon may well require different methods from those outlined here, as may speech regarded solely as a process of social interaction" (TEI Consortium, *TEI P5*).

40 TEI Consortium, *TEI P5*.
41 FADGI, "Embedded Metadata in Broadcast WAVE Files."
42 These include CIDOC's Conceptual Reference Model (CRM), the Library of Congress's Functional Requirements for Bibliographic Records (FRBR), the Dublin Core Metadata Initiative (DCMI), Contextual Ontology Architecture (COA), and the Motion Picture Experts Group rights management standard, MPEG-21.RDF (IASA). These features are also captured in TEI P5.
43 Tedlock, *Spoken Word*, 249. Tedlock's *Spoken Word* is also referenced in Johansson et al., "TEI A12 M1"; Johansson et al., "TEI A12 M2"; Sherwood, "Elaborate Versionings"; and Bernstein, *Close Listening*.
44 Chion, "Three Listening Modes," 50.
45 Bernstein, *Close Listening*, 13.
46 Sherwood, "Elaborate Versionings."
47 Sherwood, "Elaborate Versionings"; Bernstein, *Attack of the Difficult Poems*, 127.
48 Crystal and Quirk, *Systems*, 42.
49 For example, Adriana Cavarero argues for demythicizing oral performance. She critiques the viewpoint of Chion ("Three Listening Modes"), McLuhan (*Essential McLuhan*), and Ong (*Presence of the Word*), who at once essentialize the voice as "presence" and disembody and mythicize orality. Oral culture, McLuhan argues, gives "us simultaneous access to all pasts. As for tribal man, for us there is no history. All is present, and the mundane becomes mythic" (*Essential McLuhan*, 370). With this viewpoint, Cavarero asserts, we treat language as a code "whose semantic soul aspires to the universal" and render "imperceptible what is proper to the voice" ("Multiple Voices," 530).
50 FADGI, "Embedded Metadata in Broadcast WAVE Files."
51 Karpiński, "Boundaries of Language."
52 Karpiński, "Boundaries of Language," 47.
53 Schmidt, "TEI-Based Approach."
54 "An Interesting Session."
55 Tsur, *Poetic Rhythm*, 14.
56 Boas is quoted in Tsur, *Poetic Rhythm*, 51.

WORKS CITED

Bardzell, Shaowen, and Jeffrey Bardzell. "Towards a Feminist HCI Methodology: Social Science, Feminism, and HCI." In *Proceedings of the SIGCHI Conference on Human Factors in Computing Systems*, 675–84. CHI '11. New York: ACM, 2011.

Becker, S., M. Shröder, and W. Barry. "Rule-Based Prosody Prediction for German Text-to-Speech Synthesis." In *Proceedings of Speech Prosody 2006*, edited by Rüdiger Hoffman and Hansjörg Mixdorff, 503–6. Dresden: TUD Press, 2006.

Berg, Marc. "The Politics of Technology: On Bringing Social Theory into Technological Design." *Science, Technology, and Human Values* 23, no. 4 (1998): 456–90.

Bernstein, Charles. *Attack of the Difficult Poems: Essays and Inventions*. Chicago: University of Chicago Press, 2011.

Bernstein, Charles. *Close Listening: Poetry and the Performed Word*. New York: Oxford University Press, 1998.

Boas, Franz. "On Alternating Sounds." *American Anthropologist* 2, no. 1 (1889): 47–54.

Boase, S. *London-Lund Corpus: Example Text and Transcription Guide*. London: Survey of English Usage, University College London, 1990.

Bolinger, D. *Intonation and Its Parts: Melody in Spoken English*. Stanford, CA: Stanford University Press, 1986.

Bowker, Geoffrey C., and Susan Leigh Star. *Sorting Things Out: Classification and Its Consequences*. Cambridge, MA: MIT Press, 2000.

Buckland, Michael K. "Information as Thing." *Journal of the American Society for Information Science* 42, no. 5 (1991): 351–60.

Cavarero, Adriana. "Multiple Voices." In *The Sound Studies Reader*, edited by Jonathan Sterne, 520–32. New York: Routledge, 2012.

Chion, Michael. "The Three Listening Modes." In *The Sound Studies Reader*, edited by Jonathan Sterne, 48–53. New York: Routledge, 2012.

Clement, Tanya. "Distant Listening: On Data Visualisations and Noise in the Digital Humanities." *Digital Studies* 3, no. 2 (2012).

Crystal, David, and Randolph Quirk. *Systems of Prosodic and Paralinguistic Features in English*. The Hague: Mouton, 1964.

Downie, J. S., D. K. Tcheng, and X. Xiang. "Novel Interface Services for Bioacoustic Digital Libraries." In *Proceedings of the 8th ACM/IEEE-CS Joint Conference on Digital Libraries*, 423. New York: ACM, 2008.

Edwards, Jane A., and Martin D. Lampert. *Talking Data: Transcription and Coding in Discourse Research*. New York: Psychology Press, 2014.

Federal Agencies Digitization Guidelines Initiative (FADGI). "Embedded Metadata in Broadcast WAVE Files, Version 2," April 23, 2012. www.digitizationguidelines.gov/guidelines/digitize-embedding.html.

Federal Agencies Digitization Guidelines Initiative (FADGI). "About," December 10, 2010. www.digitizationguidelines.gov/about.

Feinberg, M. "Two Kinds of Evidence: How Information Systems Form Rhetorical Arguments." *Journal of Documentation* 66, no. 4 (2010): 491–512.

Frohmann, Bernd. *Deflating Information: From Science Studies to Documentation.* Toronto: University of Toronto Press, 2004.

Haraway, Donna. "Situated Knowledges: The Science Question in Feminism and the Privilege of Partial Perspective." *Feminist Studies* 14, no. 3 (1988): 575–99.

"An Interesting Session Yesterday at the National Academy of Sciences." *Washington Star,* April 19, 1878.

International Association of Sound and Audiovisual Archives (IASA). "Metadata." Accessed November 19, 2017. www.iasa-web.org/tc04/metadata.

Johansson, Stig. "The Encoding of Spoken Texts." *Computers and the Humanities* 29, no. 2 (1995): 149–58.

Johansson, Stig, Lou Burnard, Jane Edwards, and And Rosta. "TEI AI2 P1 [Spoken Texts Workgroup] Objectives and Deadlines 22 October 1990." TEI Consortium, October 22, 1990. www.tei-c.org/Vault/AI/ai2p01.tei.

Johansson, Stig, Lou Burnard, Jane Edwards, and And Rosta. "TEI AI2 M1 Minutes of Meeting Held at University of Oslo." TEI Consortium. August 9–10, 1991. www.tei-c.org/Vault/AI/ai2m01.txt.

Johansson, Stig, Lou Burnard, Jane Edwards, and And Rosta. "TEI AI2 M2 Minutes of Meeting Held at Oxford University." TEI Consortium. September 29 and October 1, 1991. www.tei-c.org/Vault/AI/ai2m02.txt.

Johansson, Stig, Lou Burnard, Jane Edwards, and And Rosta. "TEI AI2 W1 Working Paper on Spoken Texts University College London." TEI Consortium. October 1991. www.tei-c.org/Vault/AI/ai2w01.txt.

Karpiński, Maciej. "The Boundaries of Language: Dealing with Paralinguistic Features." *Lingua Posnaniensis* 54, no. 2 (2012): 37–54.

McLuhan, Marshall. *Essential McLuhan.* Edited by Eric McLuhan and Frank Zingrone. New York: Basic Books, 1995.

Mills, M. "Deaf Jam: From Inscription to Reproduction to Information." *Social Text* 28, no. 1 (2010): 35–58.

Ong, Walter J. *The Presence of the Word: Some Prolegomena for Cultural and Religious History.* New Haven: Yale University Press, 1967.

Pound, Ezra. *Polite Essays.* London: Faber & Faber, 1937.

Rooth, Matt, and Michael Wagner. "Harvesting Speech Datasets for Linguistic Research on the Web." Digging into Data Conference, National Endowment for the Humanities, Washington, DC, 2011. Accessed January 11, 2018. https://ecommons.cornell.edu/handle/1813/34477.

Schmidt, Thomas. "A TEI-Based Approach to Standardising Spoken Language Transcription." *Journal of the Text Encoding Initiative* 1 (June 2011): n.p. http://jtei.revues.org/142.

Sherwood, K. "Elaborate Versionings: Characteristics of Emergent Performance in Three Print/Oral/Aural Poets." *Oral Tradition* 21, no. 1 (2006): 119–47.

Sperberg-McQueen, M., and L. Bumarde, eds. "Guidelines for the Encoding and Interchange of Machine-Readable Texts." Draft version 1.0. Chicago and Oxford: Association for Computers and the Humanities/Association Computational Linguistics/Association for Literary and Linguistic Computing, 1990. Accessed November 19, 2017. https://quod.lib.umich.edu/cgi/t/tei/tei-idx?type=HTML&rgn=DIV2&byte=64782.

Sterne, Jonathan. "Sonic Imaginations." In *The Sound Studies Reader*, edited by Jonathan Sterne, 1–18. New York: Routledge, 2012.

Svartvik, J., and R. Quirk, eds. *A Corpus of English Conversation*. Lund, Sweden: Lund University Press, 1980.

Tedlock, Dennis. *The Spoken Word and the Work of Interpretation*. Philadelphia: University of Pennsylvania Press, 1983.

TEI Consortium, eds. *TEI P5: Guidelines for Electronic Text Encoding and Interchange*. Version 2.6.0, January 20, 2014. www.tei-c.org/Guidelines/P5.

Tsur, Reuven. *Poetic Rhythm: Structure and Performance: An Empirical Study in Cognitive Poetics*. Brighton, UK: Sussex Academic Press, 2012.

"A FOREIGN SOUND TO YOUR EAR"

Digital Image Sonification for
Historical Interpretation

MICHAEL J. KRAMER

So don't fear if you hear a foreign sound to your ear.
— BOB DYLAN, "It's Alright Ma (I'm Only Bleeding)"

INTRODUCTION: Mance Lipscomb's Silhouette

Photographs are visible, but photography is not only
a "visual" practice.
— MARGARET OLIN, *Touching Photographs*

We see a musician's back in silhouette. He sits in a chair, on an outdoor stage, facing away from view. A microphone stand rises in front of him while an acoustic guitar head, with its tuning pegs, juts out from one side of his body. You can just make out the horizontal stripes on the back of his work shirt. There is a large audience before him, sitting in steeply raked rows. In contrast to his shadowy form, their bodies are illuminated by sunlight (fig. 8.1).

The image is silent, of course; we cannot hear anything. Nonetheless, as a visual portrayal of a powerful sonic moment, it speaks volumes. Taken at

FIGURE 8.1 Mance Lipscomb performs at the Berkeley Folk Music Festival, July 1963. Photographer unknown (possibly Philip Olivier). COURTESY OF BERKELEY FOLK FESTIVAL ARCHIVE AT NORTHWESTERN UNIVERSITY SPECIAL COLLECTIONS.

a folk music festival, the photograph conveys the intense attention this lone performer commands from the crowd. The man on stage is African American songster Mance Lipscomb, a sharecropper and musician from Navasota, Texas. He performs on a beautiful summer day in 1963 at the Berkeley Folk Music Festival, which took place annually between 1958 and 1970 on the University of California's flagship campus. The photograph captures the second appearance of Lipscomb at the Berkeley festival after his debut at the 1961 event. Assisted in his journey to California by folklorist Chris Strachwitz, this working-class black man, raised under the oppressive conditions of Jim Crow segregation, appears before a primarily white, middle-class audience.[1] Lipscomb plays his bluesy acoustic songs at the Greek Amphitheater. The venue, whose construction was funded by California newspaper magnate William Randolph Hearst at the turn of the twentieth century, was modeled after the ancient open-air venue at Epidaurus. It was intended to serve as a symbol of Berkeley's aspirations to become the "Athens of the West."[2] That afternoon in the summer of 1963, at the crown jewel of California's prestigious postwar system of public higher education, in a space designed to link modern American democratic aspirations to classical antiquity, a man born to slaves took center stage.[3]

The photograph resides in the Berkeley Folk Music Festival Archive, which is housed in Northwestern University's Charles Deering McCormick Library of Special Collections and consists of over 35,000 artifacts. Currently in the process of digitization, the archive's holdings include business records, correspondence, notes, publicity materials, and much more, but the richest documentation is visual: posters, programs, and especially photographs, of which there are over 10,000.[4] This particular image captures a crucial moment of folk-revival transformation. In the click of the camera, Mance Lipscomb emerges from the shadows into the light, from the margins of society to a new place of prominence. We get to see the African American songster in the process of dissolving from one role into another—we watch the silhouette of a rural Texas sharecropper becoming a global folk music legend.

There is plenty to notice in the visual details of this photograph, but the Berkeley Folk Music Festival was, as its name suggests, a fundamentally aural event. As images such as this one go digital, can computational analysis reveal more about the sonic dimensions of the festival—and about the place of sound in historical understanding more broadly? To be sure, we cannot (at least not yet) magically recover the music being made in the instant when this photograph was taken. What we can do is move between the optic and the aural through new circuits of computational exploration to bring out concealed historical information and to generate more compelling historical interpretations. This chapter argues that through practices of digital image sonification we can expand what Fred Gibbs and Trevor Owens call "the hermeneutics of data and historical writing."[5] The digital "remediation" of the image—its passage from an earlier mode of representation into binary data—provides an opportunity to open ears as well as eyes more fully to the echoes of the past.[6] We not only can access but also experience and analyze artifacts and evidence in fresh ways to produce better history from our source materials.

Viewing a digital version of Mance Lipscomb's silhouette at the Berkeley Folk Music Festival in 1963 means that at one level we are no longer looking at the original photograph. We are instead looking at it through many removes: a digital version of a photographic print taken from a negative that used chemical processes to register light on bodies and objects in a past moment. Scanned digital images do some hard travelin', and these displacements may be troubling to the historically minded. Is the past receding from view as primary sources shift from older modes of mediation such as print and photography to the digital domain? Are we taking one more step back from the original moment in time? I contend no. Digitization does not necessarily mark a loss of access to evidence. Nor does it inevitably distort the past.[7] As this essay investigates, remediation becomes an opportunity for developing more critical thinking about the ontology—which is to say the very being—of what historical sources are and, from there, for harnessing the specific qualities of encoded digital data to foster more sensitive interpretations of history.

We should keep in mind, of course, that no artifact prior to the digital—whether it be text, sound recording, moving image, or object—offers an entirely transparent view of history. They are all mediations of one sort or another. With their odd combination of immediacy and distancing, photographs are an especially uncanny mode of representation, as commentators such as Roland Barthes and Susan Sontag have famously noted.[8] When a 1963 photograph of Mance Lipscomb moves into digital form, it becomes the newest link in an ongoing chain of representational reconfigurations stretching back to the moment in time itself. And even that moment has a medial quality in that Lipscomb's appearance at the Berkeley Folk Music Festival took place within a performance context and within a history of folk revival values, ideas, expectations, and relationships.[9] Here is not merely "raw data" to be plotted, measured, and visualized in some reductive quantitative manner, but rather a remediated representation of the past that can be processed and analyzed—both by computers and by humans—through methods made possible by its shift in underlying format to the digital domain.[10]

What is intriguing about that underlying format is that digitized photographs are more ductile, modular, and pliable in relation to other artifacts when all move into the compatible state of binary code. Computers, in this sense, are convergence machines: they bring into one unified underlying form what previously were quite different types of mediation.[11] In the digital domain, we might still speak of images, sounds, or text as distinct categories, but at the computational level they are all now bits and bytes, electronic on-and-off pulsations.

What can we do with this convergence into binary code? Among digital humanities scholars, a kind of synesthetic approach is emerging. Texts get charted, physical spaces interactively mapped, sounds graphed. The urge, however, is almost entirely to visualize data.[12] The optic dominates. Yet as a sonic event, the Berkeley Folk Music Festival asks that we also attend to the aural. We might do so by adding "sonification" to the mix alongside visualization.[13]

What follow are descriptions of three experiments with digital image sonification. Each seeks to reveal new interpretations of the Berkeley Folk Music Festival and the history of the U.S. folk music revival in the 1960s and to examine post–World War II American cultural history more broadly.[14] Taken together, they present the outlines of a hermeneutic approach to digital data that centers on shifting images into the domain of sound through their shared form as computer code.[15] First, *digital sound design* draws on practices in theater and cinema production to pair related images and sounds. These pairings, even if taken from different events, moments in time, or locations, offer new combinatory representations of the past that illuminate—amplify might be the more accurate term—historical meanings. Second, *data fusion* brings together digital data to produce a new multimedia object, and with it fresh historical knowledge. Finally, *data sonification* unleashes sounds from the data of the visual medium itself; hearing the data of an image allows one to see it differently; this expanded sensory access to evidence provides an impetus to more accurate and original historical interpretation. These three activities—digital sound design, data fusion, and direct sound sonification—remind us that the digital has the capacity to deepen our understanding of the past if we use computers inventively. In the digital medium, we can do more than just stare at Mance Lipscomb's silhouette; we can also more fully sound out its significance.

DIGITAL SOUND DESIGN: A Mount Rushmore of the Folk Revival

Sustained interpretative engagement, not efficient
completion of tasks, would be the desired outcome.
— JOHANNA DRUCKER, "Performative Materiality
   and Theoretical Approaches to Interface"

The trio of faces and upper bodies forms a kind of Mount Rushmore of the folk revival. Photographed in 1964, blues songster "Mississippi" John Hurt, Appalachian folk singer Arthel Lane "Doc" Watson, and Berkeley master of ceremonies as well as songwriter, folk singer, and professor of oceanography Sam Hinton stand together, shoulder to shoulder, backstage at the Greek Amphitheater during the Berkeley Folk Music Festival (fig. 8.2).

FIGURE 8.2 "Mississippi" John Hurt, Sam Hinton, and Arthel "Doc" Watson at the Berkeley Folk Music Festival, 1964. PHOTOGRAPH BY KELLY HART. COURTESY OF BERKELEY FOLK FESTIVAL ARCHIVE AT NORTHWESTERN UNIVERSITY SPECIAL COLLECTIONS.

There is no known audio of Hurt, Hinton, and Watson performing at the 1964 Berkeley festival; however, the three performers were making studio recordings (as well as live recordings at other venues) at the time. In digital sound design, these audio tracks can be paired up with the image in combinatory patterns that heighten our sense of the ways in which the formal details in the photograph and audio recordings relate to larger cultural contexts and interpretive ideas.

Borrowed from film, television, and theater production, concepts of sound design pay close attention to how sound *accompanies* visual representation and vice versa.¹⁶ In the digital medium, sound design offers a framework for uniting—or more precisely, collaging—previously unlinked historical images and sounds to bring them into perceptual and analytic play with one another. It harnesses a kind of "maker" approach for historical interpretation.¹⁷ To be sure, one could do much of this without digital technology: a carousel of slides and an old-fashioned cassette boom box might do the trick; so too might historical re-creations of past musical events.

These fictitiously bring "alive" the past by inventively mixing sound and images. Digital technology does not break with these approaches but rather enhances them in two ways: through intensified "versioning" that allows one to compare many different iterations of sonic and visual materials; and through the introduction of chance operations and generative possibilities derived from algorithmic manipulations.[18]

To be clear, my goal is not to join recordings of Mississippi John Hurt, Doc Watson, or Sam Hinton to the "Mount Rushmore" image of them because doing so would offer an unmediated and pure path back to the past. Instead, my efforts turn in precisely the other direction, embracing the remix as historical consciousness itself. To experiment with digital sound design is to engage with fraught but lively alignments and realignments of image and sound across impossible distances of time. It is to reassemble evidentiary elements in creative ways to better understand the past, not magically revisit it as some kind of fantastical virginal state. The many sonic and visual details of a digital sound design strike against each other synesthetically, reminding us that we only can know the past as a constellation of fragments that are always in motion, pushing and pulling on each other, producing a fecundity of interpretive truths out of their relational juxtapositions and associations as intermixed evidence.[19]

To start, let us look at the image of Mississippi John Hurt, Sam Hinton, and Doc Watson without sound. The trio stand before a wall backstage at the Hearst Greek Amphitheater in July 1964. Hurt, the blues songster from Avalon, Mississippi, looks off to his left, warmly, with the hint of a smile on his lips and his acoustic guitar clutched in the crook his right elbow and shoulder. Hinton, the professor of oceanography who served as master of ceremonies at the Berkeley Folk Music Festival, looks down with a goofy grin, his thin, striped tie stretched straight within the lapels of his tweed blazer. The blind multi-instrumentalist Doc Watson holds a banjo from a shoulder strap slung over his plaid jacket, pulling it slightly off-kilter, his hair neatly parted, but with a tuft sticking up in the back. It is an extraordinary image in its own right, conveying distinctive qualities that many in the folk revival projected onto these three famous revival performers: Hurt's softness and sweetness, Hinton's genial and endearing awkwardness, and Watson's unflappability.

Taken by festival staff photographer Kelly Hart on his Pentax camera, the photograph also, perhaps accidentally, reveals a lurking ideological urge within the 1960s folk movement: it is the dream of constituting an integrated, harmonious collective out of the fragmented and painful inequalities of race, class, age, and region in the United States, particularly the American South. With the civil rights movement reaching a crest of confrontational activity during the summer of 1964—often known as Freedom Summer after the name given to the interracial campaigns to register African American voters in the Jim Crow South—Hurt, Hinton, and Watson become a kind of symbolic string band trio, giving us the look of a more ideal America, unified in song. They do so at a festival that took place in the very same campus spaces that would soon be taken over by the influential Free Speech Movement, underway at Cal in the fall of that same year.[20]

But what was the song this symbolic string band trio was playing, exactly? How do we better hear as well as see this harmonious image of musical, racial, and regional communion? Digitization holds some possibilities. Once digitized, the image can be integrated with recordings of Hurt, Hinton, and Watson from that same period to create a digital sound design that asks the beholder to pivot between image and sound, to hear what these musicians sounded like in 1964 in relation to what they look like in the photograph, and to be able to do so in a mutating relationship of notes to visual details.

As an exercise in digital sound design, I created a collage of Hurt's version of the African American spiritual "Mary, Don't You Weep" (recorded by Peter V. Kuykendall at Wynwood Recording Studio in Falls Church, Virginia, in March 1964) and Doc Watson's version of the Dock Boggs song "Country Blues," which was released on Watson's debut album for Vanguard Records, also in 1964.[21] The process of editing the two tracks together, interspersing verses and sections into and out of one another using the free sound-editing software Audacity, caused me to pay far more careful attention to the music's content, tone, and more subtle performative dimensions (fig. 8.3).[22]

I became far more sensitive to Hurt's loping fingerpicked guitar style, so laconic yet determined, as Watson's relentless clawhammer banjo attack intrudes on it. The timbres ring out in such contrast: Hurt's thuddy, steady guitar playing compared to Watson's twangy picking, which pushes forward, clanging with urgency. Yet the two sounds are connected: when the performances were collaged, the syncopations of Hurt's melodic work on the upper strings of his guitar suddenly resembled Watson's banjo work. And the North Carolinian's great rhythmic sense, undergirding his cascades of notes, became crucial to his playing when juxtaposed against the famous, thumping tick-tock of Hurt's thumbed bassline, such a quintessential part of his sound. Here are two styles of playing stringed instruments that are quite different, they share certain qualities.[23]

The texts of each song are different, too—indeed almost opposite each

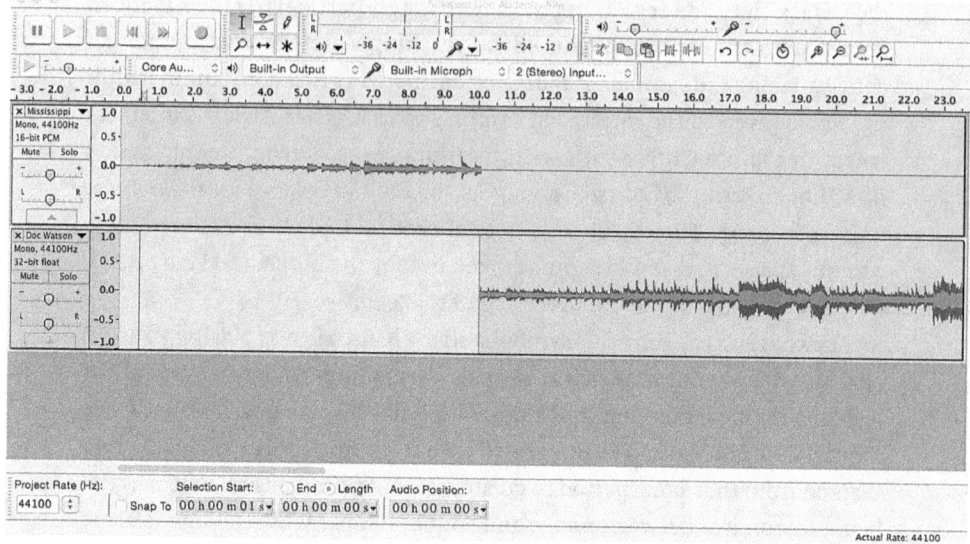

FIGURE 8.3 Remixing Mississippi John Hurt and Doc Watson in Audacity sound-editing software.

other. Hurt sings a religious hymn while Watson performs a sinner's lament. The character in "Mary, Don't You Weep" almost seems to be singing to himself, but his story is one of collective perseverance: "If I could I surely would stand on the rock where Moses stood . . . Pharoah's army got drownded, oh Mary don't you weep." The character in Watson's "Country Blues" sings to an audience but speaks of inner demons driving him to ruin: "Come all you good time people, while I've got money to spend." Hurt's words are about endurance while Watson's are about a kind of explosion of agony. Hurt's are testimonial, while Watson's are confessional. Yet the words start to intersect with each other as well, in the sense that both musicians' songs emphasize strength in the face of struggle, a refusal to look away from pain, fear, or threats of annihilation.

Stylistic musical comparison is one thing, but collaging Hurt's and Waton's respective sound recordings to then listen to them while looking at the photograph of the two men along with Sam Hinton asks us to more carefully consider how the image works as symbolic commentary both within and on the folk revival. It is, of course, just a photograph snapped backstage of these three stars of the folk scene, and yet the performers carry representational meaning and affective energy with their bodies. Because I purposely left Hinton's music out of the mix to emphasize his role as interloper and interlocutor, the digital sound design intensifies the question of how the South was represented, how its presence lurked, in the image. I had noticed the question of region in a superficial manner when simply looking at the photograph, but the more I looked and listened simultaneously, the more geographic negotiations began to ring out. The image offers a rich iconographic representation of the folk music revival's intense focus on the South even as it took place way out West. At Berkeley, in images such as this one, the desire to remix the American South's legacy of racial segregation and oppression is prominently on display. I did not see this until I heard it: collaging Hurt's and Watson's music as a soundtrack for looking at Hart's photograph revealed the regional interplay between South and West as antagonistic race relations were reimagined into new, more integrated formations.

Sound also heightened my awareness that the very bodies of the figures themselves make this symbolic racial and regional remixing possible. After all, these bodies, viewable, are most known for the sounds they made. To deliver to the image their sonic power as folk-revival performers in 1964 allows the intersections of body, music, culture, race, and region to emerge more evocatively. Sound and image together, in other words, produce a greater sense—both sensorially and semantically—of the meanings buried within the appearance of Hurt and Watson flanking Hinton. Listening to the audio collage, I suddenly noticed how these men's bodies took on archetypal demeanors (some might say stereotypical projections) of the folk-revival imagination. The camera positions Hurt, the former Mississippi farmhand, with an inner calm and deep empathy for others. There is a slight slump to his shoulders, but he is not defeated. These are shoulders that could bear weight, and did.[24] The stockiness of his chest is more pronounced, too. In life he was a small man, but not a slight one. His posture presents not weak humility so much as a strong inner reserve. Sound and image here converge: his individual poise and communal energy, the quiet, whispered quality of his singing style, seem to trace the creased wrinkles around his eyes and mouth. And another quality appears, too, one central to the reception of Hurt within the revival: his sly trickster sensibility. The edge of something more devilish below the sweetness, some kind of little, indestructible lilt, starts to dance across the surface of his gaze, turned away from the camera and toward some horizon beyond the frame of the image.[25] This photograph from 1964 does not "come to life," but the semantic and even the affective dimensions of the interaction between photographer, folk-revival milieu, and figures in the image itself do.

If Mississippi John Hurt's music and image together intensify an understanding of his appeal as an easy-going African American songster within the folk-revival imagination, Doc Watson becomes the fiercely independent Appalachian mountain man.[26] Listening to him perform while looking closely at the photograph, I began to consider his toughness, the ferocity lurking behind his friendly smile, and, most of all, the way he turned the seeming disability of his blindness into an assertion of selfhood. Paired with his performance of "Country Blues"—a rounder's testimonial of stubborn rage, pride, fury, and shame—Watson's appearance in the photograph more deeply communicates his role as a heroic figure within the folk-revival context. As with Hurt, Watson's body and sound combine to carry an entire range of associations about race, class, gender, and region. He is the white bluesman, the hearty Appalachian farmer drinking moonshine, the millworker on a bender, the gentle sage on a front porch in the mountains, all rolled into one. As a performer, Watson drew on all these projections placed upon him by folk revivalists, using them for his own expressive ends, finding his own place within—and sometimes through—their mediations.

Additional sonic experiments with the design led me to consider not only race and region but also class as a dimension of the photograph. Using Audacity, I panned the respective tracks to extreme ends of the stereoscopic spectrum, as if to echo the ways in which Hurt and Watson flank the middle-class Californian Sam Hinton. Was one end of the spectrum the Mississippi Delta, the other Appalachia? Yes, but the panning effect also made me consider the commonalities between rural black and white working-class experiences within southern life. Hurt and Watson came from different regions, but the structuring economic and class forces at work in their seemingly divergent agrarian settings were not entirely dissimilar. Delta cotton plantations and Appalachian cotton mills had many things in common, from the cotton itself to the kinds of hierarchies of power that arose from harvesting it and bringing it to market in industrialized modes of production.[27] Heard in stereo, Hurt and Watson were not only far apart; they also, as the photograph suggests, shared certain class origins that brought them into the same space. Even the very sounds they made arose from circulations of musical styles across the divisions of region and race, but not necessarily of class position, within the South.[28] The use of two-channel panning heightened my awareness not only of differences but also of similarities between the two men, particularly in contrast to Hinton as the more middle-class figure, who stands quite literally in the middle between them.

These basic efforts to bring image and sound together reaped valuable interpretive results, but computation offered additional possibilities by allowing access to the chance operations made available through algorithmic experimentation. For instance, within Audacity one can apply a sliding-time-scale/pitch-shift filter to audio data. Rather than alter the audio myself by consciously pairing audio with image, I momentarily ceded greater autonomy to the software program and its automated calculations by employing the filter. This has an air of avant-garde, John Cage compositional philosophy to it, but introducing chance into the sounds also paradoxically became a means for deeper, more precise historical scrutiny.[29] When I applied the sliding-time-scale/pitch-shift filter to my audio remix and played it while looking at the image, I started to notice issues of gender even more profoundly than I had in past viewing/listenings. My use of the filter let the computer determine when pitch rose or fell randomly in the audio of Hurt and Watson that I had created to accompany the Mount Rushmore image. (Imagine the high tones of Alvin and the Chipmunks singing followed suddenly by the basso profundo of Johnny Cash.)

The coincidences that ensued from using the filter reminded me that to alter the pitch of a singer's voice points to intensely affective, sensorial dimensions of masculinity present in musical performance. These are, as Barry Shank and others note, quite linked to assumptions about racial identity.[30] The filter raised the pitch of Hurt's singing, taking it somewhat closer in timbre and tone to the more pinched, moaning styles of other Mississippi Delta blues singers such as Robert Johnson.[31] As a contrast to his actual singing voice, which was far softer and "mellower," the altered pitch highlighted how he evoked a different kind of black masculinity within the folk-revival matrix. Coupled with the image of him in his signature button-up collared shirt and bowler hat, the algorithmically altered sound design clarified how Hurt, a man whom one folk revivalist described as a leprechaun and another as the original hippie, performed this alternative, softer style of masculine appearance.[32] It was not an essence but rather an assemblage of what we might call "glitched" details that summon gender, race, class, and region (we could add age here, too) into play with each other.[33]

The algorithmic shift raised Hurt's pitch, but since it was operating on a time scale, it lowered the sound of Watson's voice in my audio remix. His altered baritone reminded me of how his singing differed from the more famous "high lonesome" sound of bluegrass, perhaps the most famous of Appalachian-associated genres of music. To hear the algorithmically transformed voice of Watson rendered even lower brought out the ways in which he might be understood as a transitional figure with regard to questions of gender, race, and region. Within the folk revival, he was a link from the startling, high-pitched singing of someone like Bill Monroe—whose vocal style, as Robert Cantwell argues, arose from very traditional Appalachian modes of masculine identity formation—to a singing style more associated with the crooning male voice that was commonplace on radio and recordings from Watson's childhood. Typically thought of as the ultimate traditional musician who, with the encouragement of folklorist Ralph Rinzler, reached back before bluegrass to earlier mountain music styles, Watson might also be thought of as reworking traditional white southern rural and working-class masculinity into a more modern guise.[34] In place of the high lonesome sound, he sang on the lower frequencies of life at the cusp of tradition and modernity.

He did so in part by channeling into his voice an African American blues aesthetic drawn from both Piedmont and Delta traditions.[35] He also retained his interest in rockabilly and jazz, which he had been performing in a roadhouse band in North Carolina before meeting Rinzler in the early 1960s.[36] In the "Mount Rushmore" photograph, Watson's clothing suddenly takes on a new cast. Viewed to the sound of Watson's computationally lowered voice, the musician's modern-cut, green plaid suit suddenly suggests more than first meets the eye in contrast to the banjo that dangles from Watson's shoulder. This traditional musician smuggled various contemporary styles, strains, and gestures into his "old-time" sound.

In these ways, through formal manipulations accomplished both by human manipulation and algorithmic computation, digital sound design deepens the interpretive possibilities of examining images from the folk revival. Digital sound design does not bring us magically back to the past itself, for we can never make that journey. What we can do is intently make use of digital technology to notice visual and aural details more effectively as they relate to larger social and cultural forces. Constructing history through the new perceptual filters offered by digital technologies, we may develop better interpretations of the past, opening up many lines of thinking rather than narrowing analysis to one, limiting position. Digital sound design fosters an enriched clarity and precision even as it reminds us of the dense multiplicity of historical meaning present in the evidentiary record of the past, particularly when it comes to cultural expression and experience. Bringing a multisensory constellation of imagery and sounds together, digital sound design shows how our designs on the past are always shaped by the ways we—and now our computers—choose to arrange and rearrange history in the present.

DATA FUSION: A Revised *Humbead's Revised Map of the World*

> Jorge Luis Borges' story about a map . . . equal in size to the territory it represented has been re-written as a story about indexes and the data they index. Now the map has become larger than the territory.
> — LEV MANOVICH, "Database as Symbolic Form"

It is certainly a map, but the closer you look, the stranger it gets: San Francisco, Los Angeles, New York City, Cambridge, and Berkeley are the major countries; North Africa and Southeast Asia are landlocked within them; Boston is a small country with a Cape Cod–like peninsula just off the southern tip of this alternative imagining of the "geospatial" imagination. The "Rest of the World" is merely an island on the northwestern periphery, just barely bigger than Nashville. Look more closely, and the edges that frame the map contain hundreds, even perhaps thousands, of names: a wide-ranging, almost crazed list of participants in the mid-twentieth-century U.S. folk music revival. It includes everyone from Bob Dylan and Joan Baez to lesser-known local folkies to inspirational figures such as Groucho Marx. These names encircle and frame "the Great Naked Sea," out of which a sea serpent, a Poseidon-like sea king, and a large yellow bathtub duck all splash.

This remarkable cartographic fantasy, titled *Humbead's Revised Map of the World* and conceptualized by Bay Area folk scenester Earl Crabb (Humbead) and folk musician, instrument maker, and graphic artist Rick Shubb, offers new possibilities for pivoting back and forth between the visual and the aural (fig. 8.4). To borrow from the theories of cultural geography and spatial history, Shubb and Crabb transform absolute space into the relational representations of a Bay Area folkie's "mattering map."[37] Which is to say the map captures an embedded perspective on place. There is a bird's-eye view here, but what the bird sees on *Humbead's Map* reminds us that no two birds see the world below them in quite the same way. Rick Shubb remembers that the idea for the map arose in 1967 in a Berkeley music shop when Earl Crabb commented to a hitchhiker trying to get from Berkeley to Kansas City that he should put "New York" on his sign instead when standing by the side of the road looking for rides. This was because, for Crabb, "New York is closer." The humorous difference between geographical and cultural distance led Crabb and Shubb to the design of a map that reimagined the world from the perspective of a Bay Area folk music participant. Then they added the names

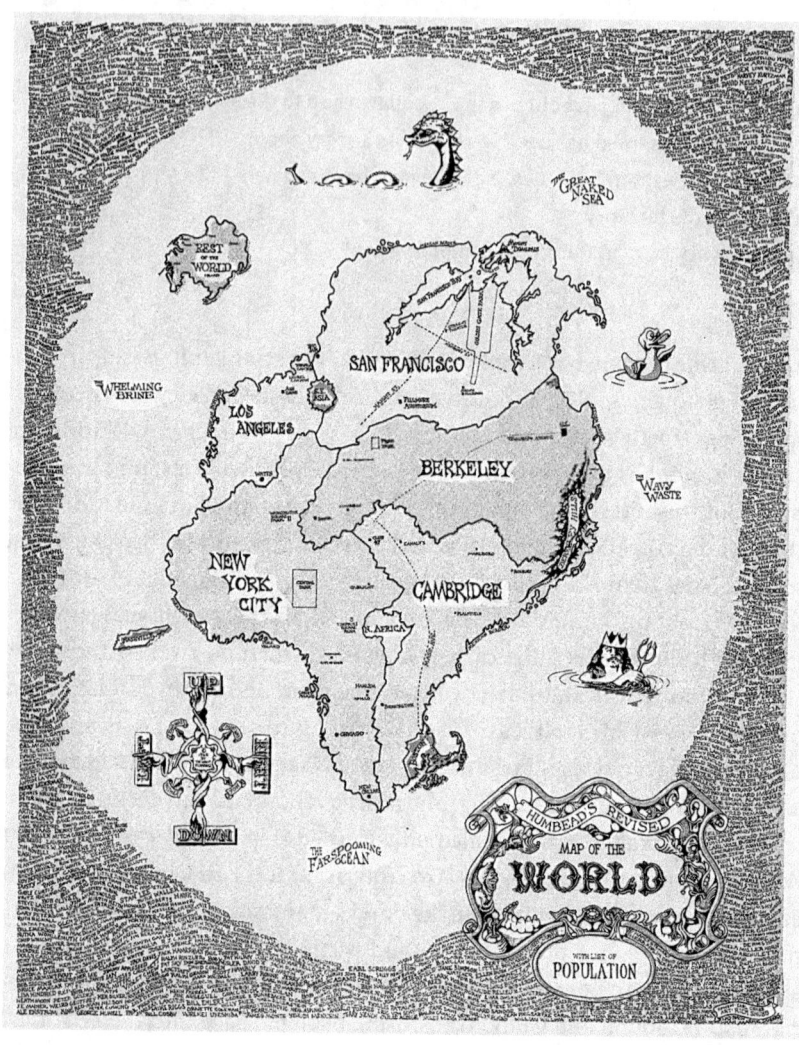

FIGURE 8.4 *Humbead's Revised Map of the World, 1968.* BY EARL CRABB AND RICK SHUBB.

around the edges, partly as a gag and also on the theory that this would inspire anyone listed to purchase a copy of the map.[38]

Sound shaped the making of the map and its content, but, of course, as a printed creation it is silent. Once brought into the digital domain, the map provides us the opportunity to address the interplay of cartographic visual representations and their sonic inspirations in one mediated space. In a version of what New Media scholar Lev Manovich calls data fusion, sound data can be layered over the digital version of the map in correspondences that thicken the contextual historicization of the Berkeley folk music scene. Manovich is thinking more of massive combinations of data made possible by computation, but his concept also works at smaller scales. It enables connections between visual and sonic information that intensify a map's implicit commentaries, in this case the tonal shadings of wit that also contain critique. These subtle gestures are at the affective root of this map about roots music. A more textured feel for the past results from remediating its source materials through the additional layer of digital data fusion. Bringing visual and sound elements together provides a more immersive experience of the map's "mapping" of the U.S. folk music revival.

Manovich defines data fusion as "using data from different sources to create new knowledge that is not explicitly contained in any of them."[39] This is what bringing sound and image together in a digital version of the map accomplishes. In other words, when fused with audio material through creative compositional choices, *Humbead's Revised Map* gets revised once again. In the process of sonification, a synesthetic interaction of visual and aural data expresses more than either medium could do individually. Whereas the digital sound design of Mississippi John Hurt, Doc Watson, and Sam Hinton did something similar by layering two media—image and sound—on top of one another, data fusion draws on far more sources and puts them into play with each other not only to see a photograph's meanings more robustly but also to produce a new kind of object born from the combination of many sources.

Using the common technology of a clickable "image map," my data fusion of *Humbead's Revised Map of the World* both accentuates and elaborates Crabb and Shubb's conceptual playfulness. It fuses audio clips with various objects, landmasses, topographical notations, names, and "nations." Many of these are straightforwardly discographic or informational—clicking on a name on the frame of the image, for instance, plays a signature song by that performer. But other sounds emphasize the affective experience of the map's cartographic commentary itself. For instance, I once again used

Audacity to create sound files at different volumes so that the size of each of the map's countries and regions correlates to the volume of the connected audio track. Bigger places in their imaginations *sound* bigger; they are also longer tracks and sometimes contain a greater amount of audio overlaid in my Audacity remixes to register the density of sonic information shaping the cartographic visual representations on *Humbead's Map*. Berkeley, San Francisco, New York City, and Cambridge play louder, longer, and more densely textured audio tracks, while Los Angeles is quieter, shorter, and simpler. Here, through sonification, I figuratively (and literally) turned up the volume on the map, using sound to intensify its cartographic choices and their implicit meanings.

Further sonifications and data fusions included placing transitional tracks at the boundaries between each "nation." At the border of New York City and Cambridge, one can select the introduction to Bob Dylan's version of "Baby Let Me Follow You Down," which mentions the "green pastures of Harvard University" as the place where he claims to have learned the song from fellow folk-scene hipster Eric Von Schmidt.[40] Recorded in New York City while Dylan was gaining acclaim in the Greenwich Village folk scene, the song's introduction is a snippet of audio that resonates with the geographic imaginings of Crabb and Shubb out on the West Coast in terms of the relationships and connections it reveals within the networks of the folk revival.

As it should, the duck in the middle of the ocean quacks.

One shortcoming of the sonification of *Humbead's Revised Map* is that it still privileges the visual over the aural: one has to see and click on the map in order to hear its sounds. In the future, I hope to create a version that reverses this orientation: it would consist of audio tracks that one selects to reveal visual information from the map. So too, the image map might become more fully interactive. The ability for users to add and further remix data would take advantage of data fusion to remind us of the social—even, potentially, the contested—nature of how the geographic and social relationships of the folk revival have been characterized. Users might be able, one day, to develop their own "mattering maps" of the folk revival, revising *Humbead's Revised Map of the World* yet again.

Overall, the effort to sonify the map not only fills in the sonic gaps but also amplifies the sensorial and affective dimensions that contributed to and were so crucial to its making and its effects. Data fusion produces a new object, and in doing so, it joins the spirit of revision already present in the original *Humbead's Revised Map*. Working at the experiential level, the distortions of the map that emerge from digital sonification can be helpful for getting a better feel for the "worldviews" that created the map. After all, the original map is itself already a kind of distortion. (And what map isn't, even as it also conveys accurate information about what it is mapping?) Moving away from a historical document by digitally manipulating it can allow one, ironically, to get closer to it. Here one can access, experience, and then contemplate the particular sounds and the sensations that undergird the ideas of Crabb and Shubb. Data fusion seeks to avoid the mistake of privileging sight and seeing over sound and hearing. Through a synesthetic fusion of the senses, it reminds us that historians often mistakenly privilege the optic, with its rationalistic associations ("seeing is believing") over the aural, with its emotional connotations ("If music be the food of love, play on"). Data fusion challenges the too-strong distinction that historians make between vision and sound and between thought and feeling, offering instead a digital method for navigating their fluidity rather than asserting one over the other.

Fusion does not render *Humbead's Revised Map* into a unified, static whole, but rather delivers the knowledge that past historical moments and movements were just that: contingent, itinerant, in motion, never complete. The data fusion is a mutating assemblage in which audio and visual components come together into a newly mediated and provisional object. Nonetheless, its dynamic characteristics make it possible to mount many arguments about what the map contains. One can begin to piece together how the small details and silenced references of sound relate to the cartographic representation as a whole—what makes it funny yet serious all at once. Data fusion provides a way to consider the evidentiary *movement* among scales of knowledge, information, sensation, and emotion that made the folk revival feel like a *movement*, a more coherent and energized social formation.

Rocketing *Humbead's Revised Map* into the digital medium of fused data, revising it yet again, the sonification also reminds us that as much as older media forms are, as media historian Lisa Gitelman has argued, "always already new," they are also the opposite: they are always already old.[41] They are embedded in the past and can still relay its meanings. Indeed, they do so precisely through the compositional reconfigurations of data fusion. Like a space-age compass, data fusion charts a way forward for navigating maps of the past.

DATA SONIFICATION: Transferring Mance Lipscomb

The relative openness of the image/sound ... create[s]
a space for shared or alternative perspectives.
— VIRGINIA KUHN AND VICKI CALLAHAN,
 "Nomadic Archives: Remix and the Drift to Praxis"

Both digital sound design and data fusion involve bringing new sounds to bear on existing images, but the possibility also arises for transforming image data itself into sound. The third and final sonification I propose can be called data sonification.[42] Working with the photograph of Mance Lipscomb taken at the 1963 Berkeley Folk Music Festival, I wondered if new interpretations of it could be rendered from transforming visual data into correlated sound outputs. In other words, could we better hear than see the photograph's staging of this particular southern, rural, African American musician's unlikely arrival onstage to play to a large crowd at a neoclassical amphitheater built into the hillside of the flagship public university of California? Would hearing this visual representation allow a historian to analyze it more perceptively? How might a different sensorial reception of the image spark new interpretive perspectives on it? What would it mean to hear an image by listening to its digital data?

The answer to these questions rests largely on the choices made in how to process and output the visual data into sonic form. What computer programmers refer to as the "architecture" of the correlation between incoming, processed, and output data becomes the key issue. In developing a particular architecture for data sonification, I began to consider what within the design and logic of visual code (pixels, color hues, grids, vectors, and even the isomorphic algorithms that make possible tactics such as shape and facial recognition) might be productively fed into certain strategies of sound synthesis (MIDI technology for instance) to generate audio tracks that originated in the visual data but took aural output form.[43] As I have previously noted, the resulting creations were not made with the intent of recovering the sounds being made when a photograph was taken. The goal is not to return miraculously and without distortion to a past reality through some legerdemain of data-manipulation magic. Instead, it is to acknowledge that we always construct history through the form and content of our source materials. These require the development of a hermeneutics, a way of interpreting.[44] What direct data sonification offers to a digital hermeneutics is a particularly intriguing way of exploring strategies for pivoting between the optic and the aural. Movement between image and sound through their now-shared ontological status as digital data synesthetically brings the eye and ear into new kinds of sensory dialogues, enhancing perceptual access to an artifact such as a photograph. From this interplay, fresh interpretations of the historical information contained within an image can arise.[45]

To begin my data sonification experiment with the image of Mance Lipscomb's silhouette at Berkeley in 1963, I wanted to know more about how data that constitute the pixels might be sonified. I adapted ideas about "glitching" from Trevor Owens's essay "Glitching Files for Understanding: Avoiding Screen Essentialism in Three Easy Steps." As Owens points out, "Digital objects are encoded information. They are bits encoded on some sort of medium. We use various kinds of software to interact with and understand those bits. In the simplest terms software reads those bits and renders them."[46] Using an MP3 recording of the "West Virginia Rag" from the Henry Reed Collection as one of his examples, Owens purposefully mismatches software applications with different file types so that one begins to visualize how the file types contain information. Most strikingly (and obviously when one thinks about it), just as a WAV audio file possesses more sonic data than an algorithmically compressed MP3 file, so too when it is viewed as a "raw" file in an image editor, the WAV file looks bigger and more spread out than the MP3 file. You can see the audio compression of sound with your eyes.

What if we move in the opposite direction to try to hear visual data? The raw file proved to be the key starting point for reversing this process and considering how instead of visualizing sound, one might sonify an image. The raw file of a photograph is sometimes known as a "digital negative" because it contains minimally processed data of a digitally created image. This file type, in other words, consists of what a digital lens translated from light and color in the world into the pixelated patterns of digital code. Outputting the photograph of Mance Lipscomb at Berkeley in 1963 as a raw data file via a text editor, I began simply by importing the file into Audacity sound-editing software (fig. 8.5). The result was not particularly useful to the human ear for listening: a solid roar of static sounded something like Lou Reed's famous album *Metal Machine Music*.[47] The experiment did not produce particularly useful results for interpretation, but it did serve as a good reminder that what we might idealize as the most direct, unmediated, and pure translation of historical data is not, when it comes to computational remediation, necessarily the most productive for generating valuable perspectives on the past. Linking visual data to sonic form requires more elaborate sound syn-

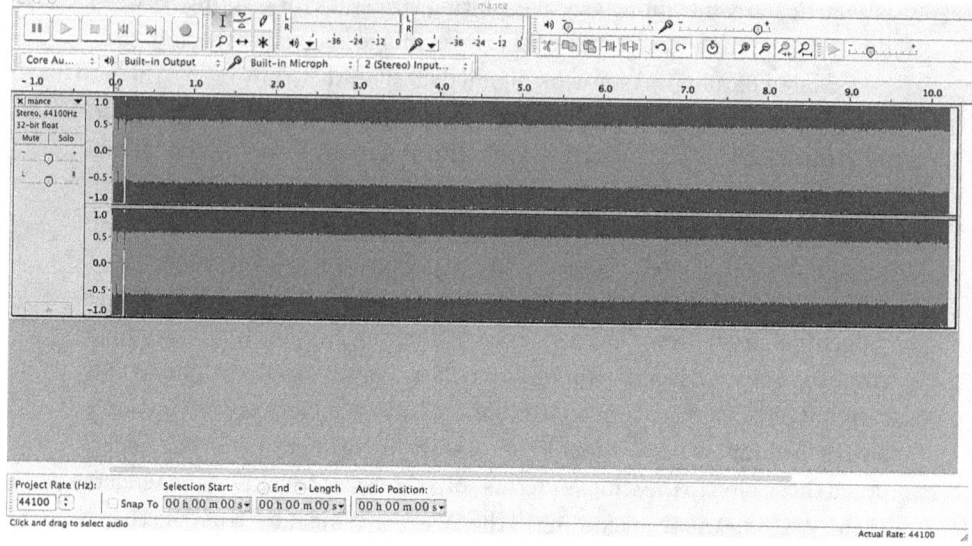

FIGURE 8.5 Mance Lipscomb at Berkeley, 1963, imported into Audacity from raw JPEG data.

thesis strategies. Sometimes you have to mess with your source materials more adventurously to grasp their significance more accurately.

My next experiment involved importing the digital image of Lipscomb into the Photosounder program designed by Michel Rouzic (fig. 8.6). Photosounder maps a JPEG file across a spectrogram whose x-axis is time and y-axis is pitch and moves by default (it is adjustable) from a low frequency of 27.5 hertz to a high frequency of 20 kilohertz. As the pixels of the image are placed within this two-dimensional Cartesian coordinate system, the intensity/brightness of each pixel is sonified through a filtered cross between white noise and pink noise: the more intense or brighter the pixel, the louder the noise.[48] What emerged from the shift to Photosounder showed that the program's sound synthesis strategy more evocatively represented the relations of color density found in the visual data than Audacity did with a raw file type.

Most fascinatingly, by offering a sonification that combined "pink noise" and "white noise" rather than generating pure sine waves, Photosounder amplified the centrality of Lipscomb's silhouetted figure. The sonification strategy produced sounds that the human ear could decipher as varying timbres, allowing one to hear densities of visual information more evocatively than one might see them. As I played the Mance Lipscomb photograph from left to right, the resulting sonification suddenly went silent when it reached Lipscomb's body at the center of the image. Then the noise, which sounded like radio static, grew louder as the sonification reached the other side of his figure and once again registered the audience pictured in the photograph.

The sonification led me to think far more carefully about the kinds of projections that folk-revival audiences at the Greek Theater enacted upon Lipscomb as he performed before them in the early 1960s. At first glance, the photograph emphasizes Lipscomb's very real arrival at center stage. He is there, present, basking in the attention of a new audience. This is a celebration. However, the Photosounder sonification suggested almost the opposite interpretation of the image: instead of placing him at center stage, it rendered him spectral and ghostly. As with Mississippi John Hurt and Doc Watson, did Lipscomb also become less a real person than a kind of symbolic keyhole through which folk revivalists thought they could unlock a whole different configuration of social relationships when it came to race, region, class, community, and the very self in modern America?

Here in the direct data sonification, the power of the photograph as a visual documentation of an aural event announces itself more clearly. The sonification reveals how Lipscomb was both extremely real to his new audience and also an enigma. He moved to the center of the folk revival, but as a silhouette, re-representing his past as an African American sharecropper and musician in rural Navisota while leaving that past behind. He achieved a whole new status as a musician and a person by carrying the shadow of that other history into broad daylight out west in Berkeley. In the click of a camera's shutter, the shadow becomes the substance, the man is his image. His silhouette, then, is not only an outline of the man but also a kind of opening, an aperture, an entry point into his larger cultural moment and the place that he, his audience, and all their history occupy within it. All the explicit and implicit negotiations, appropriations, adjustments, disorientations, alienations, connections, affiliations, and social interactions across boundaries of identity and power roar forth in the mix of photographic image and its silent silhouette of data.

Reheard through a sonic amplification of its formal qualities, the image more vividly suggests the power of the performance we glimpse in the still shot. Captured through a lens at the Greek Theater, Lipscomb captivates. The songster from Jim Crow Texas, born in 1895 as the son of former slaves, becomes a kind of king, a nobleman. He sits on a folksy wooden throne. Yet he is also small compared to the massive audience watching him. So who is ruling whom here, exactly? Perhaps what we glimpse in the image when we

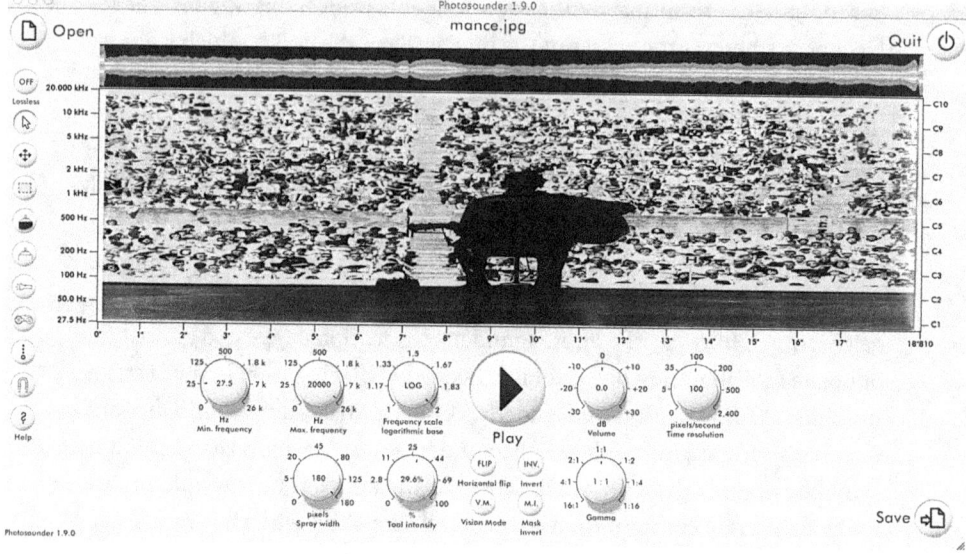

FIGURE 8.6 Mance Lipscomb at Berkeley, 1963, imported into Photosounder application.

listen to it sonified is, most of all, a transition, what folk-revival scholar Robert Cantwell calls the process of "ethnomimesis."[49] As Lipscomb traverses not only the physical but also the social distances between black working-class life in Navasota and a sunny, leisurely day on the Berkeley campus in the early 1960s, he crosses the threshold into the folk revival. He plays his past, as authentic as could be, but his performance also becomes a mimetic act of ethnic re-representation. He is performing himself in a new way. He has to. It is the only way as a performer he can reach across the divides of social location to connect with this new audience. As this cultural process occurs, we witness Lipscomb entering a different world, settling into a different position, and attaining a previously unavailable status. He does so as a charismatic performer "playing the folk," a man in the process of discovering a newfound currency gained by his virtuosic access to a fading, southern, African American vernacular musical tradition.[50] He is utterly real, but he is also, in the very same instant, a silhouette, a blank slate for the projection of fantasies, dreams, and desires by those now watching and listening to him across a chasm of social difference.

By turning this shadowy, silent photograph of Lipscomb making music at Berkeley in 1963 back into sound—a very different sound than the ones

Lipscomb made on stage that afternoon to be sure, but sound nonetheless—Photosounder helped me access a sharper, more accurate interpretive analysis of the historical moment. Even my rather technologically simple sonification was capable of doing so, and Photosounder provides ample additional opportunities for playing with the visual image data to create further sonifications of the image, including the ability to adjust frequency ranges, pixels per second, and even to manipulate pixels themselves using a set of "spray tools." The opportunities are many to experiment at the level of pixels in order to seek out their historical meaning. One can turn up the volume on certain characteristics or tone down others. In the transit between visual inputs and sonic outputs, iterative play can lead to interpretive discovery. These synesthetic collisions of image and sound are not reductive but rather generative of multiple ways of experiencing and better understanding even one historical artifact.

Other software applications and design approaches might foster additional sound synthesis strategies. For instance, one might extract semantic information rather than formal qualities from image data. Much as facial recognition software works now, isomorphic feature extraction could be employed to capture, for example, what instruments appear in the photographs of the Berkeley Folk Music Festival. How many banjos are in photographs compared to guitars? Do the ratios change over the festival's duration from 1958 to 1970? What are the various expressions that appear on Joan Baez's face as she performs onstage as compared to the many offstage portraits of her in the festival archive? Where are the eyes of audience members looking in the thousands of performance photographs? Analyzed statistically over this large data set, does gaze indicate something about what makes a particular performer charismatic or a specific concert setting enthralling for folk-revival participants? Sonification, in turn, might prove effective for perceiving undetected patterns in these kinds of semantic data extractions, for it produces a different kind of sensory experience of the data than visualization does.[51]

Overall, visual image sonification functions well on multiple scales: one can delve deeply into the meaning of the pixels in one image, or even a detail of one image, or one can go big, exploring patterns in hundreds or thousands of images. Whether it be of Mance Lipscomb's singular silhouette one summer day in 1963 or a sense of the ten thousand–plus images in the Berkeley Folk Music Festival archive, sonification enhances the perceptual and sensorial dimensions of historical inquiry by turning digital humanities analysis toward the core humanities practice of expanding the pathways by which we assess and interpret evidence. Data sonification allows us to better appreciate the multifaceted and multidimensional historical truths contained within the codification of the world into ones and zeroes.

CONCLUSION: Listening Up

Subjectivity is not merely the impure other of objectivity.
— VEIT ERLMANN, *Reason and Resonance: A History of Modern Aurality*

Moving between looking and listening by employing the tactics of digital image sonification advances conceptualizations of the digital humanities first outlined by visual theorist Johanna Drucker. She calls for the development of a "digital aesthetics" revolving around the concept of "speculative computing." Drucker's experiments at the University of Virginia's SpecLab revealed a dynamic interplay between computational experimentation and humanistic interpretation, one that seeks to expand rather than reduce meaning by attending to the pliability and ductility of digital form. Because representational objects, texts, and modes of expression are not static but rather are malleable based on positionality and perception, digital humanists, she contends, should avoid the "mathesis" of formal logic that is so prevalent in computer science. This approach, which seeks to totalize, universalize, and instrumentalize knowledge and perception, "can be challenged only by an equally authoritative tradition of aesthetic works and their basis in subjective forms of knowledge production." Drucker proposes a far more critical, qualitative, and subjective method, one that seeks to harness computation for imaginative interpretation rather than submit the critical facility of the imagination to a regime of narrow-minded quantification. As she puts it, "Neither 'works' nor 'forms' are self-evident entities. They are emergent phenomena constituted by shifting forces and fields through productive acts of interpretation."[52]

Drucker's theoretical interventions remind us that historical evidence is never transparent. We must "read" it to interpret it. We must make sense of the data for it to be meaningful. Especially with cultural material, this reading can come to include not only looking but also listening more intensively and more experimentally. To transit between the visual and the aural through their modularity as data becomes a futuristic way of journeying back into the past on speculative pathways. It is particularly powerful as a foray into the ephemeral past of events such as the Berkeley Folk Music Festival, where sonic and visual experiences intermingled to create an intangible cultural heritage worthy of scrutiny, yet easily rendered both invisible and silent. Digital image sonification becomes not so much an act of *recovering* as one of *uncovering* and *discovering*, which is to say identifying the multifaceted dimensions of historical experience.

In the end, after all, sight and sound are both grounded in the more unified experiences of sensory perception. As Jonathan Sterne points out, the separation of the senses into discrete modes is a historical phenomenon, grounded in Enlightenment thinking. The very assumption that humans have five distinct senses only emerged over time.[53] Rather than using the digital to extend and reify the separation of the senses further, we can use the flexible modularity of data to become more aware of the very history in which our senses are embedded.[54] But image sonification does not return us to some McLuhaneque unification of the senses; rather, it transits synesthetically between the optic and the aural in pursuit of meaningful explanations that arise out of the movement between the two through the mediating form of binary data.[55] As the auricular and the optic crisscross and enrich each other, a hybrid phenomenology—a way of perceiving—arises from careful attention to the ontology of digital materiality and points to what digital artifacts actually are on the material level. This approach brings us back to the rich sensorial immediacy of the past precisely by making use of its remediation—its alienation—into digital form. From there, we can produce new historical epistemologies, new ways of knowing not only what was happening but also why it mattered.

In the digital humanities, we are only just beginning to explore the possibilities of history writ in code. Like data visualization, image sonification offers one mode through which the field can push interpretation of the past forward.[56] Image sonification echoes visualization's focus on remediating and *re-presenting* data using the peculiarly modular qualities of binary code, but it also enters more uncertain territory by recalibrating the privileging of the optic over the aural in historical investigation. Digital technologies become a means for deepening comprehension not only of the visual past but also of what Sterne calls "the audible past."[57] Historians tend to be less confident about what they hear than what they see in the evidentiary record. As we enter the digital era, however, we must develop methods of accessing history through computationally mediated sources that let the noise of the future into the previously muted chambers of archival research. Through digital image sonification, we can open our ears as well as our eyes to sound in order to picture the past more completely, more accurately, more profoundly.

NOTES

The chapter title and epigraph are taken from Bob Dylan's "It's Alright Ma (I'm Only Bleeding)," on *Bringing It All Back Home*.

1. For more on Chris Strachwitz, see Rohter, "Still the Address"; Benicewicz, "Chris Strachwitz"; and Gosling and Simon's documentary film *This Ain't No Mouse Music*. For more on Mac McCormick, see Lomax, "The Collector." Strachwitz encountered Lipscomb while traveling through East Texas with fellow folklorist Mac McCormick in search of the bluesman Sam John "Lightnin'" Hopkins, whom he had similarly helped to travel to California to perform at the Berkeley festival in 1960. Strachwitz also took a photograph of Lipscomb at the 1963 festival from almost the same vantage point as the image in the Berkeley archive. It can be viewed at the Arhoolie Foundation website, http://arhoolie.org. Additionally, there is a wonderful and quite similar image in the Berkeley Folk Music Festival Collection of Lipscomb during his first appearance at the Greek Theater as part of the 1961 festival program.
2. On the history of the Greek Amphitheater, see Hyman, "UC Berkeley's Greek Theatre."
3. For more on Lipscomb, see Alyn, *I Say Me for a Parable*.
4. Berkeley Folk Music Festival Collection, Northwestern University.
5. Gibbs and Owens, "Hermeneutics of Data."
6. Bolter and Grusin, *Remediation*. See also Trettien, "Circuit-Bending Digital Humanities."
7. Indeed, what at first might seem like computational *distortions*—what digital literary scholars sometimes call deformances—can potentially lead to greater accuracy in our comprehension of the past. Mark Sample describes the term "deformance" as "a *portmanteau* that combines the words performance and deform into an interpretative concept premised upon deliberately misreading a text, for example, reading a poem backwards line-by-line." See Sample, "Notes towards a Deformed Humanities." Another digital literary scholar, Stephen Ramsay, puts it this way: "Computationally enacted textual transformations reveal themselves most clearly as self-consciously extreme forms of those *hermeneutical procedures found in all interpretive acts* [my italics]." See Ramsay, *Reading Machines*, xi. Deformance intersects with older operations that turn to stochastic methods to examine texts from new angles: everything

from Emily Dickinson's tantalizing concept of backward reading to Randall McLeod's "transformissive reading" to the systematic word-substitution tactics of the Oulipo group. Moving beyond text alone, one might also include the Zen-inflected compositional ideas of John Cage or the choreographic approaches of Merce Cunningham. To be sure, the influence of Dada, surrealism, and situationism lurks in deformance as well. See also McGann and Samuels, "Deformance and Interpretation"; Samuels, "If Meaning, Shaped Reading"; and Kramer's "Navigating the 'Screwmeneutic' Circle" and "Distorting History."

8 On the history of photography, see Barthes, *Camera Lucida*; Sontag, *On Photography*; Olin, *Touching Photographs*; and Wells, *Photography Reader*. On the representational dimensions of archival sources, see Farge, *Allure of the Archives*, and Steedman, *Dust*. For an article on the historical archive in relation to digital history, see Darnton, "Good Way to Do History."

9 On the folk revival as a milieu of cultural mediation, see Filene, *Romancing the Folk*.

10 See the essays in Gitelman, "*Raw Data*," as well as Sterne's work on formats in *MP3*.

11 Jenkins, *Convergence Culture*. See also Galloway, *Protocol*, on the technical dimensions of code that make the internet a mechanism of consolidation and control.

12 On visualization, see, for instance, Moretti, *Graphs, Maps, Trees*; Stanford University's Spatial History Project; and Staley, *Computers, Visualization, and History*.

13 The folk revival contains figures long interested in using technology to pivot between sound and sight, music and image. No less a figure than one of the founders of American ethnomusicology, Charles Seeger, father of folk-revival mainstays Pete and Mike Seeger and himself a frequent speaker at the Berkeley Folk Music Festival, supported the building of the first electronic music writer in the United States in 1956. The melograph, Seeger believed, might more accurately capture visually the sounds of world musics that did not translate well to classical Western notation. See Seeger, "Prescriptive and Descriptive," 148–95. On the melograph, see Pescatello, *Charles Seeger*, 212. The effort to notate sound is deeply connected to interest in "folk" and vernacular music; see Nettl, "I Can't Say a Thing." Thanks to Mary Caton Lingold for reminding me of this point and of Charles Seeger's important role in expanding musical visualizations beyond standard Western notation long before digital humanities scholars became obsessed with visualization. For more on technology and mediation in the folk revival, see Svec, "Folk Media," "Pete Seeger's Mediatized Folk," and "If I Had a Hammer."

14 For more on the folk revival's history and the role of festivals within that history, see Cohen, *Rainbow Quest* and *History of Folk Music Festivals*. See also Cantwell, *When We Were Good*; Filene, *Romancing the Folk*; Donaldson, "I Hear America Singing"; Lornell, *Exploring American Folk Music*; and the essays in Rosenberg,

Transforming Tradition. DeWitt, *Cajun and Zydeco Dance Music*, offers a fascinating look at the legacy of the folk revival in Northern California during more recent decades.

15 See Gibbs and Owens, "Hermeneutics of Data."
16 Documentarian Ken Burns garnered fame in part for his innovative technique of panning a movie camera across still images and photographs in his films. He often did so to a musical soundtrack. See *The Civil War*.
17 The "constructionist" philosophy of the "maker movement," with its producerist ethic of actively doing things with objects instead of passively observing them or merely critiquing them, has been influential in digital humanities. See Hatch, *Maker Movement Manifesto*; Donaldson, "Maker Movement"; and the website of the University of Victoria's Maker Lab in the Humanities (accessed January 14, 2018, http://maker.uvic.ca). For critiques of the maker movement, see Chachra, "Why I Am Not a Maker"; and Morozov, "Making It."
18 The concept of "versioning" has begun to emerge as a key tactic in digital literary studies, but less so in digital history. See Kirschenbaum, "The .txtual Condition" and "Save As." See also Manovich's notion of "variability," in *Language of New Media*, 36–45, and the UVic Maker Lab in the Humanities website article, "Versioning Modernism."
19 We are closer here to Bruno Latour's "actor-network" theory, in which objects possess agency because of their positioning within an ever-evolving matrix of social relationships. See Latour, *Reassembling the Social*.
20 Hart described his camera in an email correspondence with the author, May 30, 2017. On Freedom Summer, see McAdam, *Freedom Summer*, and Watson, *Freedom Summer*. On the Free Speech Movement at the University of California, see Cohen and Zelnik, *Free Speech Movement*, and Rorabaugh, *Berkeley at War*. Cantwell describes an image of Bob Dylan, Joan Baez, the Freedom Singers, and other folk luminaries holding hands and singing together at the closing 1963 Newport Folk Festival in these terms; see *When We Were Good*, 19, 351–52.
21 Hurt, "Mary, Don't You Weep"; Watson, "Country Blues."
22 Audacity: Multi-Track Audio Editor and Recorder software (accessed November 21, 2017, http://audacity.sourceforge.net).
23 Hurt's original track seems to have been recorded in stereo; Watson's was recorded in mono and then remastered as stereo for later re-releases. Both tracks reflect the folk-revival aesthetic of doing little technologically to manipulate the recorded sound.
24 For more on Hurt's life, see Ratcliffe, *Mississippi John Hurt*.
25 In using this term, I am linking Hurt to the scholarly literature on the African American trickster figure, particularly as a cultural innovation forged under duress due to racism, slavery, Jim Crow, and other oppressive conditions in the United States. See Baker, *Blues, Ideology*; Gates, *Signifying Monkey*; and Floyd, "African American Modernism."
26 For more on Watson's life, see Gustavson, *Blind but Now I See*.

27 See, for instance, Beckert, *Empire of Cotton*, for a history of these cross-regional structural commonalities. On southern music and class, see Malone, *Singing Cowboys* and *Don't Get Above Your Raisin'*.
28 On the circulation of southern musical styles, see Miller, *Segregating Sound*.
29 For Cage's reflections on the use of chance operations in his compositional strategies, see Kostelanetz, *Conversing with Cage*.
30 Shank, "'That Wild Mercury Sound'"; see also Radano, *Lying Up a Nation*.
31 Debates have raged among scholars about song style in the Mississippi Delta. Early blues scholars romanticized a more "primitive" and supposedly authentic, anticommercial style in the region; more recently, scholars such as Elijah Wald have pointed out that singers such as Robert Johnson performed in a wide range of styles that suited the desires of their audiences and reflected access to the range of twentieth-century American popular music; see Wald, *Escaping the Delta*.
32 Barry Olivier, director of the Berkeley Folk Music Festival, described Hurt in a conversation as a leprechaun-like figure. Hurt's manager, Boston-based photographer and folk music producer Dick Waterman, thought of him as the archetypal hippie; see Waterman, "John Hurt."
33 On glitching, see Krapp, *Noise Channels*.
34 Cantwell, *Bluegrass Breakdown*.
35 Thanks to Mary Caton Lingold for making this observation about the African American blues presence in Watson's performance style.
36 See chapter 9 of Gustavson, *Blind but Now I See*.
37 White, "What Is Spatial History?" On mattering maps, see Grossberg, *Dancing in Spite of Myself*, 13.
38 Quoted in Shubb, "Saturday Morning."
39 Manovich, *Software Takes Command*, 339, italics in original.
40 Dylan, "Baby Let Me Follow You Down."
41 Gitelman, *Always Already New*.
42 Tools of data sonification include the Photosounder application by Michel Rouzic (accessed November 21, 2017, http://photosounder.com) and TAPoR's Voyant Bubbles software (accessed November 21, 2017, www.tapor.ca/?id=11). See also "Say It with Pictures"; and Davies, Cunningham, and Grout, "Visual Stimulus."
43 Computer programmer Bill Parod of the Northwestern University Information Technology division has been immensely helpful in conceptualizing and beginning to implement this idea.
44 Gibbs and Owens, "Hermeneutics of Data."
45 Data sonification also raises fascinating questions about what counts as legitimate evidence. At one level, it distorts a photograph into a sonic form that is so removed from the original and so odd-sounding to the human ear that it might perhaps seem useless for inquiries into the past. Then again, sonified data is still linked to the original artifact; it merely remediates the data of the digital

image itself, which was already a remediation of a photograph, which in turn was a remediation of a past moment.

46 Owens, "Glitching Files."
47 Reed, *Metal Machine Music*. The album's four tracks, each precisely sixteen minutes and one second long, consist of guitar feedback manipulated at various speeds by Reed.
48 See "Photosounder Version 1.9 User Guide" (accessed January 14, 2018, http://photosounder.com/documentation.php). Michel Rouzic provided further explanation in two emails to the author, January 21, 2015.
49 Cantwell, *Ethnomimesis*.
50 For analysis of "playing the folk," see Hale, "Black as Folk"; see also Hamilton, *In Search of the Blues*.
51 The use of sound to index large datasets has been the most common approach to sonification (and visualization, for that matter) in the digital humanities. See, for instance, fascinating projects such as Joque's Listening to the Dow; Listen to Bitcoin by Laumeister; and Listen to Wikipedia by LaPorte and Hashemi. A more adventurous version can be found in Foo's ongoing Data-Driven DJ experiments. Of course, the sound of something like ten thousand banjo samples ringing out in unison seems like a wonderfully twisted thing to get to hear in its own right.
52 Drucker, *SpecLab*, xiii, 17, 23. See also Drucker, "Humanities Approaches"; and Sayers, "New Poster."
53 Sterne, *Audible Past*; Erlmann, *Reason and Resonance*; Levin, *Modernity and Hegemony*.
54 A variety of fields—visual culture studies, sound studies, and new media studies—have already begun to traverse the senses as historical forms. On visual culture studies, see Mitchell, *Picture Theory* and *What Do Pictures Want?* See also Sturken and Cartwright, *Practices of Looking*. For sound studies, see endnote 57 below. On new media studies, see Wardrip-Fruin and Montfort, *New Media Studies Reader*, and Giddings and Lister, *New Media and Technocultures Reader*.
55 See McLuhan, *Understanding Media* and *Medium Is the Massage*.
56 In this way, the practice of digital image sonification for historical interpretation offers a response to the call by archivists such as Bertram Lyons for computational scholarship to shift into a "post-digitization" phase. Rather than merely preserving artifacts digitally, we must start to figure out what we can do with them once they exist within a digital environment. See Lyons, "Editorial."
57 Sterne, *Audible Past*. The field of sound studies to which Sterne's book belongs has already begun this project of recalibration. See books such as Corbin, *Village Bells*, and Smith, *Listening to Nineteenth-Century America*. For overviews, see Sterne, *Sound Studies Reader*, and Bull and Black, *Auditory Cultures Reader*.

WORKS CITED

Alyn, Glen. *I Say Me for a Parable: The Oral Autobiography of Mance Lipscomb, Texas Bluesman, as Told to and Compiled by Glen Alyn*. New York: Norton, 1993.

Arhoolie Foundation: Preserving Vernacular Culture. Website. Accessed November 21, 2017. http://arhoolie.org.

Baker, Houston A., Jr. *Blues, Ideology, and Afro-American Literature: A Vernacular Theory*. Chicago: University of Chicago Press, 1984.

Barthes, Roland. *Camera Lucida: Reflections on Photography*. Translated by Richard Howard. New York: Farrar, Straus and Giroux, 1981.

Beckert, Sven. *Empire of Cotton: A Global History*. New York: Knopf, 2014.

Benicewicz, Larry. "Chris Strachwitz and the Arhoolie Story." Accessed January 12, 2018. www.bluesart.at/NeueSeiten/pageA54.html.

Berkeley Folk Music Festival Collection. Charles Deering McCormick Library of Special Collections, Northwestern University.

Bolter, Jay David, and Richard Grusin. *Remediation: Understanding New Media*. Cambridge, MA: MIT Press, 2000.

Bull, Michael, and Les Black, eds. *The Auditory Cultures Reader*. New York: Bloomsbury, 2004.

Burns, Ken, dir. *The Civil War*. 1990. Arlington, VA: PBS.

Cantwell, Robert. *Bluegrass Breakdown: The Making of the Old Southern Sound*. Champaign-Urbana: University of Illinois Press, 1984.

Cantwell, Robert. *Ethnomimesis: Folklore and the Representation of Culture*. Chapel Hill: University of North Carolina Press, 1993.

Cantwell, Robert. *When We Were Good: The Folk Revival*. Cambridge, MA: Harvard University Press, 1996.

Chachra, Debbie. "Why I Am Not a Maker." *Atlantic*, January 23, 2015. www.theatlantic.com/technology/archive/2015/01/why-i-am-not-a-maker/384767.

Cohen, Ronald D. *A History of Folk Music Festivals in the United States*. Lanham, MD: Scarecrow Press, 2008.

Cohen, Ronald D. *Rainbow Quest: The Folk Music Revival and American Society, 1940–1970*. Amherst: University of Massachusetts Press, 2002.

Cohen, Robert, and Reginald E. Zelnik, eds. *The Free Speech Movement: Reflections on Berkeley in the 1960s*. Berkeley: University of California Press, 2002.

Corbin, Alain. *Village Bells: Sound and Meaning in the Nineteenth-Century French Countryside*. Translated by Martin Thom. New York: Columbia University Press, 1998.

Darnton, Robert. "The Good Way to Do History." *New York Review of Books*, January 9, 2014. www.nybooks.com/articles/archives/2014/jan/09/good-way-history.

Davies, Gareth, Stuart Cunningham, and Vic Grout. "Visual Stimulus for Aural Pleasure." Accessed November 21, 2017. https://pdfs.semanticscholar.org/5bba/fd5a3e75bdd4ad08975a91ebd0141073b51e.pdf#page=23.

DeWitt, Mark F. *Cajun and Zydeco Dance Music in Northern California: Modern Pleasures in a Postmodern World.* Oxford: University Press of Mississippi, 2008.

Donaldson, Jonan. "The Maker Movement and the Rebirth of Constructionism." *Hybrid Pedagogy*, January 23, 2014. www.hybridpedagogy.com/journal/constructionism-reborn.

Donaldson, Rachel Clare. *"I Hear America Singing": Folk Music and National Identity.* Philadelphia: Temple University Press, 2014.

Drucker, Johanna. "Humanities Approaches to Graphical Display." *Digital Humanities Quarterly* 5, no. 1 (2011): n.p. www.digitalhumanities.org/dhq/vol/5/1/000091/000091.html.

Drucker, Johanna. "Performative Materiality and Theoretical Approaches to Interface." *Digital Humanities Quarterly* 7, no. 1 (2013): n.p. www.digitalhumanities.org/dhq/vol/7/1/000143/000143.html.

Drucker, Johanna. *SpecLab: Digital Aesthetics and Projects in Speculative Computing.* Chicago: University of Chicago Press, 2009.

Dylan, Bob. "Baby Let Me Follow You Down." *Bob Dylan.* Columbia Records CS 8579, 1962.

Dylan, Bob. "It's Alright Ma (I'm Only Bleeding)." *Bringing It All Back Home.* Columbia Records CS 9128, LP, 1965.

Erlmann, Veit. *Reason and Resonance: A History of Modern Aurality.* Cambridge, MA: MIT Press, 2010.

Farge, Arlette. *The Allure of the Archives.* 1989; reprint, New Haven, CT: Yale University Press, 2013.

Filene, Benjamin. *Romancing the Folk: Public Memory and American Roots Music.* Chapel Hill: University of North Carolina Press, 2000.

Floyd, Samuel A. "African American Modernism, Signifyin(g), and Black Music." In Samuel A. Floyd, *The Power of Black Music: Interpreting Its History from Africa to the United States*, 87–99. New York: Oxford University Press, 1995.

Foo, Brian. Data-Driven DJ. Website. Accessed November 21, 2017. https://datadrivendj.com.

Galloway, Alexander. *Protocol: How Control Exists after Decentralization.* Cambridge, MA: MIT Press, 2006.

Gates, Henry Louis, Jr. *The Signifying Monkey: A Theory of African-American Literary Criticism.* New York: Oxford University Press, 1988.

Gibbs, Fred, and Trevor Owens. "The Hermeneutics of Data and Historical Writing." In *Writing History in the Digital Age*, edited by Jack Dougherty and Kristen Nawrotzki, 159–72. Ann Arbor: University of Michigan Press, 2013. http://writinghistory.trincoll.edu/data/gibbs-owens-2012-spring.

Giddings, Seth, and Martin Lister, eds. *The New Media and Technocultures Reader.* New York: Routledge, 2011.

Gitelman, Lisa. *Always Already New: Media, History and the Data of Culture.* Cambridge, MA: MIT Press, 2006.

Gitelman, Lisa, ed. *"Raw Data" Is an Oxymoron.* Cambridge, MA: MIT Press, 2013.

Gold, Matthew K., ed. *Debates in the Digital Humanities.* Minneapolis: University of Minnesota Press, 2012.

Gosling, Maureen, and Chris Simon, dirs. *This Ain't No Mouse Music: The Story of Chris Strachwitz and Arhoolie Records.* 2013. New York: Kino Lorber.

Grossberg, Lawrence. *Dancing in Spite of Myself: Essays on Popular Culture.* Durham, NC: Duke University Press, 1997.

Gustavson, Kent. *Blind but Now I See: The Biography of Music Legend Doc Watson.* Tulsa, OK: Blooming Twig Books, 2011.

Hale, Grace Elizabeth, "Black as Folk: The Folk Music Revival, the Civil Rights Movement, and Bob Dylan." In Grace Elizabeth Hale, *A Nation of Outsiders: How the White Middle Class Fell in Love with Rebellion in Postwar America,* 84–131. New York: Oxford University Press, 2011.

Hamilton, Marybeth. *In Search of the Blues.* New York: Basic Books, 2008.

Hatch, Mark. *The Maker Movement Manifesto: Rules for Innovation in the New World of Crafters, Hackers, and Tinkerers.* New York: McGraw-Hill, 2013.

Hurt, Mississippi John. "Mary, Don't You Weep." *Worried Blues.* Piedmont Records PLP 13161, 1964; re-released as *Worried Blues 1963,* Rounder Records CD 1082, 1991.

Hyman, Carol. "UC Berkeley's Greek Theatre Turns 100 Years Old This Month." September 11, 2003. www.berkeley.edu/news/media/releases/2003/09/11_greek.shtml.

Jenkins, Henry, ed. *Convergence Culture: Where Old and New Media Collide.* New York: New York University Press, 2006.

Joque, Justin. Listening to the Dow. Website. Accessed November 21, 2017. http://vimeo.com/23965023.

Kirschenbaum, Matthew G. "Save As: Michael Joyce's Afternoons." In Matthew G. Kirschenbaum, *Mechanisms: New Media and the Forensic Imagination,* 159–212. Cambridge, MA: MIT Press, 2012.

Kirschenbaum, Matthew G. "The .txtual Condition: Digital Humanities, Born-Digital Archives, and the Future Literary." *Digital Humanities Quarterly* 7, no. 1 (2013): n.p. www.digitalhumanities.org/dhq/vol/7/1/000151/000151.html.

Kostelanetz, Richard. *Conversing with Cage.* 1987; 2nd ed., New York: Routledge, 2003.

Kramer, Michael J. "Distorting History (To Make It More Accurate)." April 3, 2016. www.michaeljkramer.net/cr/distorting-history-to-make-it-more-accurate.

Kramer, Michael J. "Navigating the 'Screwmeneutic' Circle." May 2, 2012. www.michaeljkramer.net/cr/navigating-the-screwmeneutic-circle.

Krapp, Peter. *Noise Channels: Glitch and Error in Digital Culture.* Minneapolis: University of Minnesota Press, 2011.

Kuhn, Virginia, and Vicki Callahan. "Nomadic Archives: Remix and the Drift to

Praxis." In *Digital Humanities Pedagogy: Practices, Principles, and Politics*, edited by Brett D. Hirsch, ch. 12 (n.p.). Cambridge, UK: Open Book Publishers, 2012. http://openbookpublishers.com/htmlreader/DHP/chap12.html.

LaPorte, Stephen, and Mahmoud Hashemi. Listen to Wikipedia. Website. Accessed November 21, 2017. http://listen.hatnote.com.

Latour, Bruno. *Reassembling the Social: An Introduction to Actor-Network-Theory*. New York: Oxford University Press, 2005.

Laumeister, Maximillian. Realtime Bitcoin Transaction Visualizer (formerly Listen to Bitcoin). Website. Accessed November 21, 2017. www.bitlisten.com.

Levin, Michael, ed. *Modernity and the Hegemony of Vision*. Berkeley: University of California Press, 1993.

Lomax, John Nova. "The Collector: Mack McCormick's Huge Archive of Culture and Lore." *Houston Press*, November 19, 2008. www.houstonpress.com/2008-11-20/news/the-collector-mack-mccormick-s-huge-archive-of-culture-and-lore/full.

Lornell, Kip. *Exploring American Folk Music: Ethnic, Grassroots, and Regional Traditions in the United States*. Jackson: University Press of Mississippi, 2012.

Lyons, Bertram. "Editorial." *Journal of the International Association of Sound and Audiovisual Archives* 43 (July 2014): 2–4.

Maker Lab in the Humanities, University of Victoria. "Versioning Modernism," September 2, 2012. http://maker.uvic.ca/mod.

Malone, Bill C. *Don't Get Above Your Raisin': Country Music and the Southern Working Class*. Urbana: University of Illinois Press, 2002.

Malone, Bill C. *Singing Cowboys and Musical Mountaineers: Southern Culture and the Roots of Country Music*. Athens: University of Georgia Press, 1993.

Manovich, Lev. "Database as Symbolic Form." *Convergence* 5, 80 (June 1999): 80–99.

Manovich, Lev. *The Language of New Media*. Cambridge, MA: MIT Press, 2001.

Manovich, Lev. *Software Takes Command*. New York: Bloomsbury, 2013.

McAdam, Douglas. *Freedom Summer*. New York: Oxford University Press, 1988.

McGann, Jerome, and Lisa Samuels. "Deformance and Interpretation." *New Literary History* 30 (Winter 1999): 25–56.

McLuhan, Marshall. *Understanding Media: The Extensions of Man*. 1964; reprint, Cambridge, MA: MIT Press, 1998.

McLuhan, Marshall, with Quentin Fiore, coordinated by Jerome Agel. *The Medium Is the Massage: An Inventory of Effects*. New York: Random House, 1967.

Miller, Karl Hagstrom. *Segregating Sound: Inventing Folk and Pop Music in the Age of Jim Crow*. Durham, NC: Duke University Press, 2010.

Mitchell, W. J. T. *Picture Theory: Essays on Verbal and Visual Representation*. Chicago: University of Chicago Press, 1994.

Mitchell, W. J. T. *What Do Pictures Want? The Lives and Loves of Images*. Chicago: University of Chicago Press, 2005.

Moretti, Franco. *Graphs, Maps, Trees: Abstract Models for Literary History*. New York: Verso, 2007.

Morozov, Evgeny. "Making It." *New Yorker*, January 13, 2014. www.newyorker.com/magazine/2014/01/13/making-it-2.

Nettl, Bruno. "I Can't Say a Thing until I've Seen the Score: Transcription." In Bruno, *The Study of Ethnomusicology: Thirty-One Issues and Concepts*, 74–91. 1983; revised, Urbana: University of Illinois, 2005.

Olin, Margaret. *Touching Photographs*. Chicago: University of Chicago Press, 2012.

Owens, Trevor. "Glitching Files for Understanding: Avoiding Screen Essentialism in Three Easy Steps." *The Signal: Digital Preservation*, November 5, 2012. http://blogs.loc.gov/digitalpreservation/2012/11/glitching-files-for-understanding-avoiding-screen-essentialism-in-three-easy-steps.

Pescatello, Ann M. *Charles Seeger: A Life in American Music*. Pittsburgh: University of Pittsburgh Press, 1992.

Radano, Ron. *Lying Up a Nation: Race and Black Music*. Chicago: University of Chicago Press, 2003.

Ramsay, Stephen. *Reading Machines: Toward an Algorithmic Criticism*. Urbana: University of Illinois Press, 2011.

Ratcliffe, Philip R. *Mississippi John Hurt: His Life, His Times, His Blues*. Oxford: University Press of Mississippi, 2011.

Reed, Lou. *Metal Machine Music*. RCA CPS2-1101, 1975.

Rohter, Larry. "Still the Address of Down-Home Sounds." *New York Times*, November 28, 2010. www.nytimes.com/2010/11/28/arts/music/28arhoolie.html.

Rorabaugh, W. J. *Berkeley at War: The 1960s*. New York: Oxford University Press, 1989.

Rosenberg, Neil V. *Transforming Tradition: Folk Music Revivals Examined*. Urbana: University of Illinois Press, 1993.

Sample, Mark. "Notes towards a Deformed Humanities." May 2, 2012. www.samplereality.com/2012/05/02/notes-towards-a-deformed-humanities.

Samuels, Lisa. "If Meaning, Shaped Reading, and Leslie Scalapino's way." *Qui Parle* 12, no. 2 (2001): 179–200.

Sayers, Jentery. "New Poster: Humanities on the Z-Axis." June 24, 2013. http://maker.uvic.ca/zposter.

"Say It with Pictures." *Electronic Musician*, September 1, 2008.

Seeger, Charles. "Prescriptive and Descriptive Music-Writing." *Musical Quarterly* 44, no. 2 (April 1958): 148–95.

Shank, Barry. "'That Wild Mercury Sound': Bob Dylan and the Illusion of American Culture." *Boundary 2*, no. 29 (Spring 2002): 97–123.

Shubb, Rick. "Saturday Morning." Excerpt from unpublished manuscript, 69–73. Accessed November 21, 2017. http://shubb.com/cafe/archive/Humbead.pdf.

Smith, Mark M. *Listening to Nineteenth-Century America*. Chapel Hill: University of North Carolina Press, 2001.

Sontag, Susan. *On Photography*. New York: Farrar, Straus and Giroux, 1977.

Spatial History Project, Stanford University. Website. Accessed November 21, 2017. http://web.stanford.edu/group/spatialhistory/cgi-bin/site/index.php.

Staley, David J. *Computers, Visualization, and History: How New Technology Will Transform Our Understanding of the Past*. Armonk, NY: M. E. Sharpe, 2013.

Steedman, Carolyn. *Dust: The Archive and Cultural History*. New Brunswick, NJ: Rutgers University Press, 2002.

Sterne, Jonathan. *The Audible Past: Cultural Origins of Sound Studies*. Durham, NC: Duke University Press, 2003.

Sterne, Jonathan. *MP3: The Meaning of a Format*. Durham, NC: Duke University Press, 2012.

Sterne, Jonathan., ed. *The Sound Studies Reader*. New York: Routledge, 2012.

Sturken, Marita, and Lisa Cartwright. *Practices of Looking: An Introduction to Visual Culture*. New York: Oxford University Press, 2001.

Svec, Henry Adam. "Folk Media: Alan Lomax's Deep Digitality." *Canadian Journal of Communication* 38, no. 2 (2013): 227–44.

Svec, Henry Adam. "If I Had a Hammer: An Archeology of Tactical Media from the Hootenanny to the People's Microphone." PhD diss., University of Western Ontario, 2013.

Svec, Henry Adam. "Pete Seeger's Mediatized Folk." *Journal of Popular Music Studies* 27, no. 2 (2015): 145–62.

Trettien, Whitney. "Circuit-Bending Digital Humanities," *Diapsalmata*, January 18, 2013. Republished in revised form in *Between Humanities and the Digital*, edited by Patrik Svensson and David Theo Goldberg, 181–92. Cambridge, MA: MIT Press, 2015.

Wald, Elijah. *Escaping the Delta: Robert Johnson and the Invention of the Blues*. New York: Amistad, 2004.

Wardrip-Fruin, Noah, and Nick Montfort, eds. *The New Media Studies Reader*. Cambridge, MA: MIT Press, 2003.

Waterman, Dick. "John Hurt: Patriarch Hippie." *Sing Out!*, February/March 1967, 4–7.

Watson, Arthel "Doc." "Country Blues." *Doc Watson*. Vanguard Records VSD-79152, 1964.

Watson, Bruce. *Freedom Summer: The Savage Season of 1964 That Made Mississippi Burn and Made America a Democracy*. New York: Viking, 2010.

Wells, Liz, ed. *The Photography Reader*. New York: Routledge, 2004.

White, Richard. "What Is Spatial History?" Spatial History Lab working paper, February 1, 2010. www.stanford.edu/group/spatialhistory/cgi-bin/site/pub.php?id=29.

AUGMENTING MUSICAL ARGUMENTS

Interdisciplinary Publishing Platforms
and Augmented Notes

JOANNA SWAFFORD

From the beginning of my PhD program, I knew that I wanted to examine the intersections of music and poetry. I quickly realized that print media was not the ideal format for that examination. When analyzing a musical setting of a Victorian poem for a final paper, I painstakingly included annotated excerpts of a score in the appendix and a CD so that my professor could both see and hear the musical effects I was elucidating. Unfortunately, the professor was unable read music, so the score served no useful purpose, and he found it difficult and cumbersome to associate the musical passages I described in my essay with those on the CD. Without a way to immediately unite score and audio, I knew my arguments would continue to be unintelligible to my audience. Thanks to the training I received from Networked Infrastructure for Nineteenth-Century Electronic Scholarship (NINES) and the Scholars' Lab digital humanities fellowships at the University of Virginia, I began the process of building Augmented Notes, a tool to help make the highly specialized language of music accessible to nonmusicians.

Before I started building my own tool, I first surveyed other solutions to this problem in publishing, both in print and online. In the 1960s and 70s ethnomusicologists often included LPs with their monographs so readers could hear the music the book described.[1] In the 1980s musicologists often replaced vinyl with cassette tapes.[2] By the 1990s monographs and textbooks often included CDs.[3] However, these solutions require readers to go to the extra trouble of finding the exact measures of the song on the external audio files, and this additional step reduces the likelihood that anyone will actually follow the argument by listening to the music.

Recently, some monographs have started incorporating supplemental websites to better address this problem. The second edition of Mark Katz's *Capturing Sound: How Technology Has Changed Music* replaced the first edition's CD with a continuously updated website that includes the audio and video files mentioned in the book with cross-referenced page numbers.[4] While this website enables readers to hear the music and audio in question, it still does not help readers find the exact musical phrases mentioned in articles, and those with less musical expertise will be left out of the conversation entirely.

Textbooks and print journals have followed suit: as a supplement to their anthologies of British literature, Broadview Press features a password-protected webpage that includes a section titled "Sounds of British Literature," which contains only recordings of sixteenth- and seventeenth-century songs.[5] While it is vital for literary scholars to address musical settings of songs, it is equally important to provide the score as well as the audio. The Norton anthologies for music also have elaborate websites, which include "Listen For" tutorials. These tutorials have short videos for selected songs that feature one-sentence voiceovers explaining the importance of the excerpt, followed by labeled annotated timelines of the score's structure, which become filled by a black bar as the audio track progresses through the song. While this method provides a detailed, guided tour through portions of particular songs, it does not include images of the score itself, which makes it less useful for my purposes. The *Journal of the American Musicological Society* has perhaps the best model—when users click on an article with audio, the journal displays the score excerpt and plays the corresponding audio—but even this framework is limited: the journal is primarily distributed in print, and accessing the web framework, which is protected through a paywall, is cumbersome.[6] Additionally, although the audio and score are presented together, the score is not highlighted in time with the audio: this omission is not a problem for musicologists (the primary audience for JAMS), but it would present a problem to readers who are not music specialists.

In recent years the digital sound studies community has produced a growing number of multimedia archives, but these generally opt to privilege either the score or the audio. For example, the English Broadside Ballad Archive (EBBA) at the University of California, Santa Barbara, is bringing musical settings to the fore by digitizing almost eight thousand ballads from England, and it includes facsimiles, transcriptions, and, when available, audio recordings of the ballads.[7] It also contains some essays and visualizations (including graphs and maps) of different aspects of the songs, making it an excellent resource that furthers the goals of sound studies. However, since printed ballads often included only the words and not the scores, this site reproduces that publishing strategy, sidestepping the problems inherent in making musical scores legible to nonmusicians.[8]

Recently, scholars and programmers have tried to address this problem by finding new publishing strategies to incorporate music in academic articles, as SoundCite and Scalar have demonstrated.[9] SoundCite is a tool that lets users embed sound clips in websites by following three easy steps. It enables users to place the audio file in line with the text, and even overlapping with it, so that clicking on a phrase will begin playing an audio file of that phrase. This tool is incredibly useful for publishing articles online if users are only concerned with linking audio to text, but it does not support linking audio to score images, and it sometimes glitches on mobile devices and when used with WordPress. Scalar is a publishing framework that lets users annotate media and superimpose those annotations, so users can add links and text to appear in any audio or video file. Again, however, users cannot synchronize an audio file and score, so any included scores will still be illegible to nonmusicians. Other attempted strategies include interactive CDs designed to guide newcomers to classical music through some canonical works, MIDI plug-ins for web browsers, and flash-based animated scores.[10] Some musicology periodicals have opted for an exclusively digital form to embed MP3s or YouTube videos into their analyses.[11] However, none of these projects feature good-quality audio integrated with a score, and many rely on outdated technology. The other popular option involves using video-editing software to create short film clips by manually synching scores with audio files and then posting the clips on YouTube as animated scores. While this strategy can produce an end result similar to what I had envisioned, video-editing software can be quite expensive and cumbersome, and I was looking for a tool designed to bring score and sound together, rather than one that could be rigged to do the job.

After surveying the available digital tools and projects, I concluded that nothing existed that suited my needs: I wanted to build a new publishing framework to combine audio, score, and analytical commentary in which every measure of a score would be highlighted in time with the music from an audio file. I was fortunate to have been a member of the first Praxis Program cohort with the Scholars' Lab at the University of Virginia, a graduate fellows program that gives students an intensive education in digital humanities. From that experience, I learned enough programming and basic web development to build Songs of the Victorians, an archive and analysis of Victorian song settings of contemporaneous poems.[12] Each archive page includes a recording of a Victorian song synced with its first-edition printing so that every measure of the song is highlighted in time with the music. Songs of the Victorians also has article-length analyses of these songs, which explain how the musical settings function as interpretations of their lyrics. The articles use fragments of these integrated scores as excerpts to support my analyses of the gender politics of each song: wherever I elucidate a phrase of the song, I supplement the analysis with a hyperlink (a speaker icon) that, when clicked, reveals the corresponding excerpt in which the score is highlighted in time with the audio. For example, when discussing political activist, marriage reformer, and composer Caroline Norton's best-selling song "Juanita" (1853), I use this excerpting framework to show how a musical allusion makes the song into a subtle critique of marriage rather than an endorsement of it. The melody for the first four bars of the chorus (on the words "Nita, Juanita, Ask thy soul if we should part") is the same as the melody from Handel's aria "Lascia ch'io pianga" from the opera *Rinaldo* (1711), in which an imprisoned woman laments her fate and dreams of freedom. I use the excerpting framework to play the Handel excerpt and Norton excerpt side by side so users can hear the similarities and then more readily believe my argument: that Norton used this allusion to suggest that marriage is a type of imprisonment and that Juanita, like Norton herself, wishes for freedom from a husband.

This framework made possible my interdisciplinary scholarship: without Songs of the Victorians, I would have been unable to convince my readers that women musicians could use these disarmingly simple songs, often performed in the parlor as part of a courtship ritual, to unsettle the gendered status quo, from queering the heteronormative space of the parlor to taking greater agency in courtship to critiquing marriage laws. This digital publishing framework enables literary scholars without musical experience to follow arguments that, when presented simply as musical notes printed on a page, were completely inaccessible. It has also been invaluable in conference

FIGURE 9.1 Box-drawing page of Augmented Notes.

presentations: multiple scholars have informed me that they had previously been intimidated by interdisciplinary arguments involving music, but the highlighting framework gave them new confidence in and understanding of such arguments.

The framework was so popular that I had requests to build similar sites for other scholars' conference presentations, archives, books, and articles. As I could not build customized sites for everyone, I created Augmented Notes, a generalized public humanities tool that allows users to integrate an audio file with a score to use in both academic arguments and digital archives.[13] Augmented Notes takes audio files and score images and combines them into webpages where each measure of the score is highlighted in time with the audio so that everyone, regardless of musical literacy, can follow along. It is simple to use and eliminates the need for users to understand programming. After uploading audio and image files, users are taken to a page where they click and drag to draw boxes around each measure. Users can change the size and position, numbering, and alignment of boxes (or delete them, as well). Once the entire score has been appropriately highlighted, users can proceed to the time editing page.

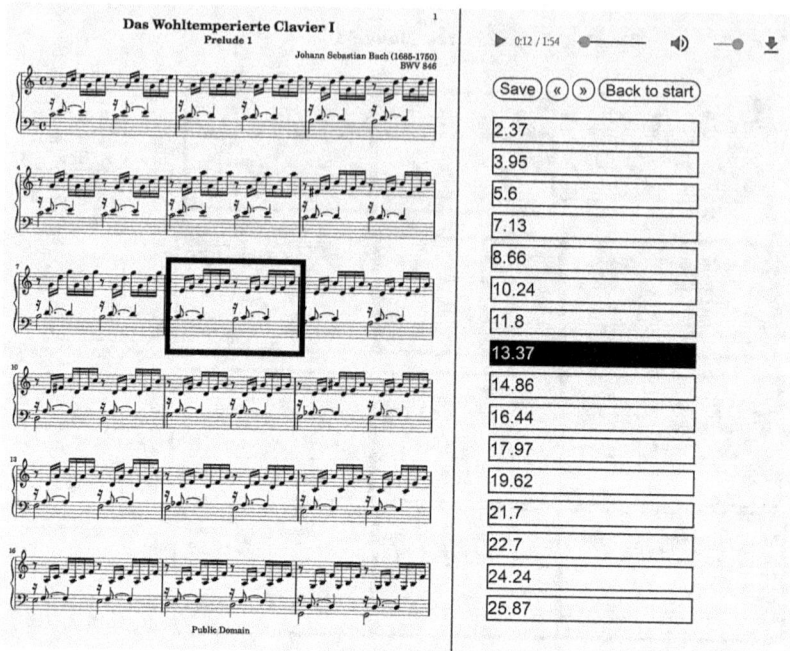

FIGURE 9.2 Time-editing page of Augmented Notes.

On the time editing page, users annotate each measure with the time when that measure ends in the audio: here, a scholar listens while the audio plays and hits a button as the audio reaches the end of each measure, thus marking the measure boundaries, which records the timestamp in the input boxes on the right-hand side of the screen. Users continue recording timestamps until the entire score has been processed. They can edit any of the timestamps, jump forward and backward by measure, and go back to the start to observe whether they properly aligned the audio with the score. Once the user is satisfied with the measure locations and times, Augmented Notes exports the measure and time information necessary to highlight each measure of the song in time with the music. Users then download a zip file with the HTML, CSS, and JavaScript files necessary for an integrated archive page, which they can then restyle themselves.

Augmented Notes also has a sandbox where users who would like to experiment with the technology but do not themselves have the requisite files can try it out.[14] This demo page includes the score and audio to Bach's Prelude No. 1 in C major (BWV 846). This song perfectly encapsulates Augmented

FIGURE 9.3 Customizable output of Augmented Notes.

Notes, because the score and audio are in the public domain (echoing the tool's open-source policies) and also because this piece opens Bach's *Das Wohltemperierte Klavier*, an influential series of keyboard pieces composed to show off the advantages of a new system of keyboard tuning. Since *Das Wohltemperierte Klavier* was Bach's attempt to unite the written and the aural while using a new technology, it nicely parallels the purpose of Augmented Notes, which is also a new system for combining the written and the aural.

Augmented Notes is already being used by scholars for archival projects, such as Romantic-Era Lyrics from the University of South Carolina and Sounding Tennyson from the University of Cambridge. It also has pedagogical uses: professors can use the interactive scores to teach or test basic score-following in music appreciation classrooms or to highlight particular motives buried in orchestral scores in more advanced classes. Because it is free, open source, and usable by anyone with computer access, scores, and audio files, it can preserve any musical cultural record and can be used for scholarly or nonscholarly purposes. It was voted first runner-up in the DH awards 2013 competition in the "best DH tool or suite of tools" category.[15]

I am currently designing options for greater user customization: specifically, I want to support highlighting musical units other than measures (including individual notes) and to allow users to draw shapes other than rectangles around the music. Additionally, I plan to build greater support for MEI (Music Encoding Initiative). Developed by Perry Roland at the University of Virginia, MEI is a type of XML for the scholarly encoding of music, just as TEI is designed for the scholarly encoding of text.[16] It is quickly becoming the standard markup for scholarly digital editions of scores. With MEI scholars can encode a measure of music in a method similar to encoding a section of a poem in TEI: they can mark up the measure, staff, chord duration, notes, and instrumentation as one might mark up a stanza, lines within a stanza, the rhyme scheme, and the speaker. Although Augmented Notes can import bar line positions and timestamps from MEI, it does not yet export MEI, and this additional functionality would increase the tool's interoperability and usefulness in the greater digital sound studies community. For instance, MEI capabilities would let me partner with Edirom, a music editor that enables the collation of music marked up in MEI so that users can easily compare multiple performances of the same song.

Although currently the site only produces archive pages, I am expanding its functionality to accommodate an excerpting framework. After syncing the score with the audio, users will be able to select the starting and ending points for different excerpts and the caption for these excerpts. The output will include a second HTML file that contains the excerpts, labeled and listed in order, onto which users can add the surrounding analytical text. I have already used this strategy in Songs of the Victorians: when the commentary discusses a particular measure, the users can click on an icon of a speaker to highlight the relevant measures of the score in time with the audio so they can hear for themselves the effect the commentary describes.

I was able to learn the HTML, CSS, and JavaScript necessary for Augmented Notes only because of my graduate school training. I became a fellow with NINES early in my PhD program, during which I learned TEI and the importance of archives and developed enough basic web development skills to build the prototype for Songs of the Victorians. NINES granted me a scholarship to attend the Digital Humanities Summer Institute (DHSI), where I took "Multimedia: Design for Visual, Auditory, and Interactive Electronic Environments," an intensive weeklong training program that prepared me for my next experience: I was one of six students in the first Praxis Program cohort at the Scholars' Lab at the University of Virginia. We had weekly meetings in which we learned best practices of digital humanities, including designating credit, drafting project charters, project management, web design, and basic programming. We worked closely with the Scholars' Lab staff, and each Praxis student developed specialties—including coding, design, and project management—in addition to expanding our generalist knowledge, ensuring that the developers could communicate clearly with the designers at all times. We were also an interdisciplinary group, so we gained experience collaborating with others outside our academic fields. Over the course of the year, we worked together to build a tool—Prism—for analyzing and comparing crowdsourced interpretations of text, and the following year the Praxis cohort added to it. As lead developer on the project, I got a crash course in JavaScript and CoffeeScript, which helped me refine the code underlying Songs of the Victorians, and my newfound knowledge of best design practices inspired me to completely redo the interface for the site. It was during that year that I came up with the idea for Augmented Notes, and the project management skills I acquired during Praxis enabled me to break the project into its smaller components and stay on task building it during my year as a Scholars' Lab fellow.

The fellows program enables up to three students each year to work closely with the Scholars' Lab staff on a research project of their choice and pays enough to reduce their teaching loads for at least a semester so they can focus on completing the project by the end of the year. With the advice and training of the Scholars' Lab, I further revamped Songs of the Victorians and created Augmented Notes. The three fellowship programs gave me the support, knowledge, and funding required for my project: without their help, I would have been unable to build my site and therefore unable to pursue my dissertation topic. Since I was writing a full-length dissertation in addition to building two digital projects, their support was particularly invaluable. Because their support enabled me to build my interdisciplinary projects, it also led first to my junior faculty position as Assistant Professor of Interdisciplinary and Digital Teaching and Scholarship at SUNY New Paltz, and later, to my position as the Digital Humanities Specialist at Tufts University. Augmented Notes, and graduate training in digital humanities more generally, truly shaped my career trajectory.

As my own history shows, new media enable new arguments: without Augmented Notes to unite audio with score, I would have had to abandon my arguments about performances of Victorian songs, as my ideas needed a new structure that incorporated sound to be legible to nonspecialists. In fact, given the rise of digital publishing options for sound studies, evidenced not only by this collection but also by the rise of companion websites for textbooks and monographs and other standalone websites and tools, it is not a surprise that sound studies is experiencing a resurgence. However, as more publishers begin to explore options for digitally representing sound, we must make sure that the digital tools that make such work possible are not limited to people with the prestige and clout to access them: rather, we should continue to produce and improve the open-source and open-access tools that make our scholarship possible for everyone, from graduate students and independent scholars to endowed chairs. Since our scholarship is often limited by our technology's ability to represent the art we analyze, we must continue to challenge our current publishing models to create the scholarship our art deserves.

NOTES

1 Søgård Jørgensen's *Qavaat* and Edström's *Sámisk musik* are just two examples of this trend.
2 For example, musicologist Nicholas Temperley's special edition of *Victorian Studies* included a cassette tape with the songs discussed in the articles.

3 This common technique is seen in such works as Rice's *May It Fill Your Soul*, Watkins's *Proof through the Night*, and Turino's *Music as Social Life*.
4 Hosted by University of California Press, the companion website for Katz's book is at www.ucpress.edu/go/capturingsound.
5 Pinch and Bijsterveld's *The Oxford Handbook of Sound Studies* also has a companion website with links to websites, MP3 files, and YouTube videos that are referenced in the text, but they do not include any transcriptions or scores; Boretz, Morris, and Rahn's *Perspectives of New Music* also includes a digital appendix.
6 Saavedra, "Carlos Chávez's Polysemic Style."
7 English Broadside Ballad Archive.
8 Douglass and Burwick's Romantic-Era Songs project, which contains popular settings of Romantic poems, adopts a similar strategy: it includes introductory materials for each song or set of songs, audio files (MP3s), and the occasional transcription of the words, but the omission of scores hampers a truly interdisciplinary analysis.
9 SoundCite, from Northwestern University Knight Lab (accessed February 28, 2015, http://soundcite.knightlab.com); and Scalar, from the University of Southern California (accessed February 28, 2015, http://scalar.usc.edu/scalar).
10 MIDI Sheet Music (accessed November 27, 2017, http://sourceforge.net/p/midisheetmusic/wiki/Home) and Sibelius Scorch (accessed February 28, 2015, www.sibelius.com/products/scorch/index.html) are MIDI-based, whereas Variations, developed at Indiana University (accessed February 28, 2015, http://variations.indiana.edu/use/index.html), is the most sophisticated flash-based approach.
11 For example, *Inbhear: Journal of Irish Music and Dance* (accessed February 28, 2015, www.irishworldacademy.ie/inbhear) and *Echo: A Music-Centered Journal* (accessed February 28, 2015, www.echo.ucla.edu) incorporate MP3s and videos.
12 You can visit the website at www.songsofthevictorians.com.
13 See Augmented Notes: A Tool for Producing Interdisciplinary Music and Text Scholarship (accessed November 27, 2017, www.augmentednotes.com).
14 For the sandbox, see www.augmentednotes.com/example.
15 "Digital Humanities Awards."
16 For more on MEI, see McIntire Department of Music: Music Encoding Initiative (accessed November 27, 2017, http://music.virginia.edu/mei). MEI's better-known counterpart, MusicXML, is another XML for music designed mainly for formatting music in composition programs such as Sibelius (accessed November 27, 2017, www.avid.com/sibelius) and Finale (accessed November 27, 2017, www.finalemusic.com). MEI enables scholars to show "areas where multiple readings or realizations of the musical content—drawn from different sources—are possible, or encode information indicating that a different hand was used to write a section, or even a particular symbol, of a manuscript. Multiple media may also be related to the encoding, providing

methods of associating audio recordings or scanned images with the musical content." Roland, Hankinson, and Pugin, "Early Music and the Music Encoding Initiative," 610.

WORKS CITED

Black, Joseph, ed. *The Broadview Anthology of British Literature*. Vol. 5, *The Victorian Era*. 2nd ed. Toronto: Broadview Press, 2012.

Boretz, Benjamin, Robert Morris, and John Rahn, eds. *Perspectives of New Music*. Accessed February 28, 2015. www.perspectivesofnewmusic.org.

Cook, Nicholas. *Beyond the Score: Music as Performance*. Oxford: Oxford University Press, 2014.

"Digital Humanities Awards: Highlighting Resources in Digital Humanities." DH Awards 2013. Accessed May 12, 2015. http://dhawards.org/dhawards2013/results.

Douglass, Paul, and Frederick Burwick, eds. *Romantic-Era Songs*. Accessed February 28, 2015. www.sjsu.edu/faculty/douglass/music/index.html.

Edström, Karl-Olof. *Sámisk musik i förvandling*. Stockholm: Caprice Records, 1988.

English Broadside Ballad Archive. Accessed February 28, 2015. http://ebba.english.ucsb.edu.

Griffiths, Eric. *The Printed Voice of Victorian Poetry*. Oxford: Clarendon Press, 1989.

Katz, Michael. *Capturing Sound: How Technology Has Changed Music*. Berkeley: University of California Press, 2010.

Pinch, Trevor, and Karin Bijsterveld, eds. *The Oxford Handbook of Sound Studies* companion website. Oxford University Press, 2011. Accessed May 13, 2015. http://global.oup.com/us/companion.websites/9780195388947.

Rice, Timothy. *May It Fill Your Soul: Experiencing Bulgarian Music*. Chicago: University of Chicago Press, 1994.

Roland, Perry, Andrew Hankinson, and Laurent Pugin. "Early Music and the Music Encoding Initiative." *Early Music* 42, no. 4 (2014): 605–11.

Saavedra, Leonora. "Carlos Chávez's Polysemic Style: Constructing the National, Seeking the Cosmopolitan." *Journal of the American Musicological Society* 68, no. 1 (2015): 99–150.

Scott, Derek B. *The Singing Bourgeois: Songs of the Victorian Drawing Room and Parlour*. 2nd ed. Burlington, VT: Ashgate, 2001.

Søgård Jørgensen, Christian. *Qavaat: Musik og Fortællinger fra Sydgrønland*. Copenhagen: ULO, 1982.

Swafford, Joanna. Songs of the Victorians. Accessed January 26, 2017. www.songsofthevictorians.com.

Temperley, Nicholas, ed. "Music in Victorian Society and Culture." Special issue, *Victorian Studies* 30, no. 1 (1986).

Turino, Thomas. *Music as Social Life: The Politics of Participation*. Chicago: University of Chicago Press, 2008.

Watkins, Glen. *Proof through the Night: Music and the Great War*. Berkeley: University of California Press, 2003.

Wood, Jennifer. "Noisy Texts: How to Embed Soundbytes in Your Writing." Burnable Books blog, edited by Bruce Holsinger, October 17, 2013. http://burnablebooks.com/noisy-texts-how-to-embed-soundbytes-in-your-writing-a-guest-post.

IV

POINTS FORWARD

10

DIGITAL APPROACHES TO HISTORICAL ACOUSTEMOLOGIES

Replication and Reenactment

REBECCA DOWD GEOFFROY-SCHWINDEN

In Paris, on the first Wednesday of every month, air raid siren tests blare from noon to 12:10. Goose bumps cover my arms every time I hear this sound. A complex set of experiences and knowledge causes my body to respond in such a way: the historical knowledge of World War II, an understanding of the siren tests' cultural context, a personal relationship with a man who grew up in Paris during the war, my imagination of living under the threat of bombings, and my current fear of terrorism. I grew up in a small town of two thousand residents in northeastern Pennsylvania, where every night at nine a fire siren wails, sounding remarkably similar to the Parisian air raid siren. Yet this Pennsylvania siren evokes quite a different response from me. As a child, upon hearing it I would drop whatever occupied me at the moment, race to the staircase of my home, and proceed to play "race-you-upstairs" against my mother. The comforting nine o'clock whistle signaled bedtime through a combination of family tradition and local timekeeping. It still conjures in me an emotional sense of safety and love. Despite the material similarity of these two siren sounds, it is a distinct set of historical, cultural, local, and personal meanings that grants each its discrete significance.

Humans experience sound through a complex set of physical, psychological, emotional, and affective processes. The siren sound is not merely an acoustic phenomenon; it is also a portion of my sonic knowledge, an internalized admixture of sensory experiences and subjective interpretations. As listeners bounce new sounds off past experiences to (re)create meaning, they simultaneously establish personal archives of sonic knowledge and a collective, social archive, as well. I was surprised to find that a quick Google search of "Paris air raid sirens" yields dozens of reflections on this aural experience, particularly by nonnative Parisians, who share my affective reaction to the eerie sound. This relational process of listening and hearing underlies what ethnomusicologist Steven Feld has called acoustemologies, or acoustic epistemologies: "the agency of knowing the world through sound."[1] Feld continually highlights how sensuous local knowledge constitutes acoustemologies; he writes, "The world sonified is the world known, the world felt, the world performed."[2] Humans know, feel, and perform the world in historically and culturally specific ways, in both individual and collective capacities.

The growth of sound studies has fueled an initiative to approach the distant past through sound. Scholarship across disciplines has begun to investigate how a sonic framework can reveal new insights into history. The first step to this research is often a material reconstruction of historical soundscapes, and scholars have turned toward digital technologies to replicate a diversity of sounds that until recently were considered lost.[3] Currently, this digital methodology tends to stop at the material. Scholars of audible history have hardly experimented with the soft side of digital media, which offers promising formats to exhibit the meanings that have been attached to sounds throughout history. Digital approaches to these more subjective facets of sound might transform acoustic reconstructions from distant sonic artifacts to intimate sonic experiences, and also challenge how scholars traditionally present their primary and archival sources. Digital formats not only simulate the process by which sentient beings constitute acoustemologies in everyday life, but they can also uncover the methodological and theoretical implications of academic publication formats. A turn to diverse media in the presentation of audible history will encourage a vital rethinking of the performance of archival research as well as scholarly production and reception. This turn might force scholars to rethink the underlying assumptions of their work, while also inviting broader audiences into the reenactment of historical acoustemologies.

Recording, Recovery, Replication, Reenactment

Despite its lack of sound recordings, the pre-recording-technology archive does not lack a historical record of sound.[4] Like Eugene Smith's ever-running loft tape, brittle documents, chipped objects, and crumbling architecture uniquely capture sound as a testament to life, life that sounded even before mechanical reproduction could fully capture it. But accusations of inauthenticity and anachronism may await digital projects representing such artifacts through twenty-first-century technologies. Past debates within history and musicology, two of audible history's parent fields, forewarn such scholarly derision. A prevalent distrust of reenactment—the re-creation of historical events in the present—pervades the discipline of history, where the practice has a popular rather than critical connotation.[5] Conversely, the field of music known as historically informed performance (HIP) uses period instruments, past performance techniques, and contemporaneous treatises to re-create in minute detail historically accurate performances of pre-recording-technology compositions. After disciplinary growing pains in the 1980s and 1990s, critics today rarely bother to condemn the authenticity of HIP scholars' research, which is generally considered, at least among musicologists, as commensurate to a historian interpreting archival documents and artifacts.[6] Somehow, historical performance practitioners dodge the lowbrow label of musicological "reenactors." The predecessors of audible history, history and musicology, have the potential to cast a shadow over still-nascent digital approaches to the field. Although an entire subdiscipline of music is dedicated to the precise replication of historical music, the re-creation of sound more generally might amount to mere philistine reenactment. Lest we forget, the other roots of audible history, by way of sound studies, are grounded in anthropology and ethnomusicology, the two closely related disciplines that gave birth to Feld's concept of acoustemologies. The dependence of anthropology and ethnomusicology on field recordings engenders sound as a serious object of study and mode of analysis. This intellectual breakthrough is unlikely to have taken place without the possibility of recording technology, just as HIP's obsession with historically faithful performances is unlikely to have developed without the possibility of infinitely repeatable musical recordings. Technology that has facilitated such fruitful methods should not be dismissed as somehow foreign to historical research.[7]

Nonetheless, methods of audible history that privilege recording technology inappropriately would indeed emphasize the twenty-first-century values that equate sound preservation with physical capture. Nonsonic media provide alternative modes of listening and hearing that might produce fresh perspectives and complement the points of audition located in speakers or headphones. Rather than favor any one technology to re-create so-called authentic sounds, an acknowledgment of how all technologies mediate sound—from musical notation to MP3s—will prove more generative. A critical stance toward the role of recording in audible history does not preclude digital methods; to the contrary, it calls for them. Digital methods grant diverse means of interaction with historical materials through formats that simulate how auditioning subjects acquired sonic knowledge in the past. The mobilization of diverse media in the pursuit of historical sounds will open up the possibilities of the pre-recording-technology archive, just as recording technology revolutionized research in anthropology, ethnomusicology, and musicology.

A lack of traditional sonic records such as field recordings in the archive inevitably concerns scholars of digital audible history. Earwitness accounts provide a bulk of pre-recording-technology sources, including reports of complaints about noise, writings by travelers who recount new acoustic surroundings, and descriptions of musical performances and other sonic events. Earwitnesses not only provide descriptive source material but also explanations of how people listened to and interpreted sound. Images, too, offer vital sources and interpretations of sound, depicting objects, technologies, spaces, and living beings that all contributed to the acoustics and soundscape of a particular time and place. Contemporaneous material culture and architecture contribute vital information for the recovery of soundscapes through acoustic and architectural modeling. Scholars of audible history may feel anxiety about the forms their sources take in comparison to sound recordings; even so, they boast a trove of evidence from which to recover ostensibly lost sounds.

Acoustics, Soundscapes, Audition

Similar to its textual counterpart, digital audible history scholarship tends to fall within three overlapping categories: acoustics, soundscapes, and audition. The three categories build on one another, with soundscapes serving as a vertex where scientific acoustics and subjective audition meet. Acoustics reconstruct the sound of historical spaces quite accurately with the help of acoustic and architectural modeling, while soundscapes push beyond acoustics to include the objects and beings that populate historical spaces. Audition then incorporates not only histories of listening but also histories of the corporeality of hearing and the cerebral processes by which humans attach meaning to sound.[8] In the footsteps of R. Murray Schafer, who first articulated the concept of historical earwitness accounts, scholars have built on his work to consider the subjectivity of such sources—like my opening anecdote—and by extension, the subjectivity of the auditory.[9] Of course, the two avenues of sounding environment and auditioning subject are rarely discrete, and the most compelling audible histories often lie at their intersection.[10] The distinction between acoustic space as material and quantifiable and soundscape as listener-centric has been refined in the field of archaeoacoustics, which generally considers acoustic space as an entity that can be modeled and analyzed and soundscape as constructed, at least in part, through listeners' experiences.[11] Thus, although soundscapes can be partially reconstructed through acoustics, the listener's agency in the constitution of soundscapes remains crucial—an assertion that rests at the heart of Feld's definition of acoustemologies, as well. A brief review of two digital audible history projects illustrates these generalizations about the nascent field.

English professor John N. Wall led the cross-disciplinary team of researchers (from North Carolina State University; Cambridge, Massachusetts; and London) who created the Virtual Paul's Cross Project, "A Digital Re-creation of John Donne's Gunpowder Day Sermon." The team used architectural modeling and acoustic simulation software to reconstruct the performance of John Donne's Gunpowder Day sermon, which was scheduled to take place at the St. Paul's Cathedral Churchyard on November 5, 1622, but took place indoors instead due to inclement weather.[12] Using digital tools typically employed to predict how sound will interact with space, the Virtual Paul's Cross Project combines architectural, environmental, performative, and social factors to reconstruct the experience of Donne's famous speech. In addition to considering the architectural and acoustic features of the churchyard space, project participants also worked closely with experts on seventeenth-century oration and John Donne. The project culminated with an installation at the North Carolina State campus in Raleigh, which immersed visitors in a 270-degree, wraparound image of St. Paul's Churchyard, as reconstructed by visualization software, while audio clips from twenty-one speakers broadcast what listeners might have experienced during Donne's outdoor sermons. Because the churchyard was later destroyed by fire in 1666 and the speech never took place outdoors, the project replicates the performance context as it could have been rather than as a precise replica of a sonic event from history. This approach evades the issue of historical authenticity and shifts focus instead to acoustic factors in the reception of an improvisational performance genre that existed before recording technology. The team specifically chose the Gunpowder Day sermon because it is one of the few Donne sermons to be transcribed soon after its delivery. The project goal was not to perfectly replicate a specific Donne speech but to synthesize acoustic and historical evidence toward a replication of the seventeenth-century London soundscape in which listeners experienced Donne's many sermons.

The project website offers practical, methodological, and theoretical considerations that informed the team's research, as well as sample audiovisual clips from the installation. The clips place visitors in eight different points of audition throughout the churchyard and feature varying crowd sizes from five hundred to five thousand listeners. For example, site visitors can choose to experience how the sermon would have sounded from behind the preaching station in a crowd of twelve hundred people. Along with the speech read by a specialist of seventeenth-century oration, listeners hear crowd noise, birdcalls, and dogs barking—the acoustic environment that might have surrounded Donne's performances and that would have constituted the acoustic environment of courtyard sermons. The website contextualizes this soundscape through textual descriptions of seventeenth-century sermon practices, Donne's personal style of speech delivery, and the historical and political climate of the Gunpowder Day sermon.

This project contributes an intellectually rigorous reconstruction of a historical soundscape to the field of digital audible history and achieves a variety of scholarly goals.[13] The project creators summarize the outcomes of their endeavor as a reestablishment of the relationship between lost structures and spaces, a presentation of quantitative information that corrects interpretive earwitness accounts, a simulation of how a nonrecorded performance took place in space through time, and a demonstration of how place and space affect both communication and performance.[14] The project coordinators conclude that the final product has made "the Paul's Cross sermon the subject of reflection precisely as a situated experience, a communal and participatory experience, unfolding interactively in real time and in a specific place, under specific conditions of weather, season, and urban environment."[15] Ideally, this would be the accomplishment of any digital audible history endeavor—to use sound as a way to historically and culturally situate a sonic experience or context for modern listeners.

The challenge remains to bring modern listeners into this experience as active earwitnesses to the courtyard sermons. Indeed, the project website admits that the installation presents only one side of an interactive historical performance, and the project's insistence on "correcting" seventeenth-century accounts seems to distrust the experience conveyed by earwitnesses. Research on oration practices and acoustics permits an informed replica of Donne's sermon soundscape; however, a reenactment approach to the soundscape would further guide modern listeners to consider the historical auditioning subject within the replication. The "Donne Interacting" page of the project website begins to work toward such engagement through descriptions of how Donne would have interacted with his audience, as well as how particular aspects of his speech and its acoustic context would have stimulated audience participation.[16] Although the project team considers the demonstration of how the sermon unfolded "interactively" as one of its achievements, the form of the Virtual Paul's Cross Project prompts twenty-first-century listeners to passively absorb seventeenth-century audition through textual marginalia rather than active sensory engagement.

Historian Emily Thompson's "The Roaring Twenties: An Interactive Exploration of the Historical Soundscape of New York City" offers an example of how digital audible history scholarship might encourage modern listeners to actively engage with historical audition.[17] The online project presents primary sources and archival materials about noise complaints and violations in early-twentieth-century New York City. The historical materials include everything from official noise complaints submitted to city governance to newsreels that portray the cacophonous New York City soundscape. Since recording technologies were at the cusp of booming during this period, the project also includes some contemporaneous films and sound recordings, although most materials are textual. Historical documentation of what New Yorkers considered noise—a contingent social category—situates these auditioning subjects in their historical acoustemology. In contrast to the Virtual Paul's Cross Project, which emphasizes the replication of a soundscape, "The Roaring Twenties" focuses primarily on how people heard, interpreted, contributed to, and reacted to their local soundscape.

Although Thompson's project is rooted in an era during which recording technology began to flourish, it offers insight into potential digital approaches to pre-recording-technology archives. The design of the project, produced in collaboration with web designer Scott Mahoney, offers three modes to explore the historical materials gathered by Thompson: sound, space, and time. Through the "sound" mode, visitors can sift through the materials categorically by types of sound: for example, sounds of traffic, transportation, or the home. If visitors choose to engage the materials through "space," the documents and clips appear charted on a 1933 map of New York City. Through "time," the materials are plotted chronologically on a timeline. Online visitors can choose their preferred mode to navigate through Thompson's vast amount of historical materials, or they can engage the materials from a variety of perspectives—categorically, spatially, and temporally—to acclimatize themselves to the acoustemology of 1920s New York City. The materials presented in the digital project were presented in a more traditional academic format in Thompson's book *The Soundscape of Modernity*.[18] Rather than interpreting materials for a reader and presenting them in a predetermined, written format, the digital project grants visitors the agency to choose an approach and to participate in the work of historicizing sound. By implicating the visitor in both the replication and reenactment of sonic artifacts, Thompson achieves a sophisticated balance among acoustics, soundscapes, and audition that does not excessively depend on recordings. Unlike the Virtual Paul's Cross Project, which foregrounds replication, Thompson's approach insists upon the historicization of sound. Digital audible history here is not merely a replication of sounds but also a reanimation of historical acoustemologies. The online format grants modern listeners an opportunity to explore materials that attest to audition, mimicking the experiential, nonlinear process by which humans accrue sonic knowledge in reality. As social practices, hearing and listening are constituted across two axes—physical experience of material reality and psychological interpretation of those physical experiences.[19] In dialogue, the Virtual Paul's Cross Project and "The Roaring Twenties" exemplify how digital audible history can transform historical audition for modern listeners from a mere sonic event into a sonic experience through both replication and reenactment.

Digital Approaches to Historical Acoustemologies

My contribution to the web collection "*Provoke!*, Organs of the Soul: Sonic Networks in Eighteenth-Century Paris" takes an audition-focused approach to digital audible history.[20] Published on Scalar, a born-digital, open-access scholarly publishing platform, "Organs of the Soul" connects descriptions, depictions, and transcriptions of sound found in archival and primary source documents from eighteenth-century Paris through thematic narrative pathways and subject tags. Visitors can choose to follow the paths "voice," "music," or "sound" throughout the project, or they can browse materials by more detailed tags, such as *Encyclopédie*, Rousseau, or popular song. The project streamlines diverse media on eighteenth-century Paris available across the internet—from digitized document collections on scholarly websites such as Gallica to musical recordings and videos on social media like YouTube. Connections among these sonic artifacts aim to demonstrate how sound performed, transmitted, and created knowledge in eighteenth-century Paris. Many of the pages are narrated in my own voice to make transparent the position of historians as only one of many mediators in the construction of audible history and to bring audition from textual marginalia and into the haptic experience of the project.

By allowing sonic artifacts to interact with one another, "Organs of the Soul," like "The Roaring Twenties," attempts to re-create a web of sonic knowledge that would have constituted the historical acoustemology of auditioning subjects in eighteenth-century Paris. The "Organs of the Soul" paths begin with excerpts from the *Encyclopédie*, a contemporaneous publication that offers widely accepted definitions of various subjects in eighteenth-century France. To imagine the sound of a voice through eighteenth-century French ears, one must first understand how contemporaries would have defined it, and so the voice path, for example, begins with a mid-eighteenth-century French definition of voice. The project reveals how many definitions of "voice" coexisted during this period, including a firm distinction between "the people's voice" (a consensus) versus "the public voice" (inarticulate noise of the masses). This information nuances our understanding of an earwitness account that describes a swelling public voice. Although our twenty-first-century sensibilities might interpret such a description as a positive, democratic sentiment, the *Encyclopédie* definition elucidates that the phrase actually describes popular complaints as insignificant babble.

It is challenging to convey the historical and cultural specificity of sonic experience in a way that invites audiences to actively engage with soundscape replicas. The siren anecdote that opens this chapter demonstrates the necessity of combining two digital approaches—on the one hand, the scientific reconstruction of acoustics and soundscapes, and on the other, contextualization of auditioning subjects within soundscapes. A problem of quietness persists in the "Organs of the Soul" project. It does not recreate as much as it attempts to describe the sonic reality of eighteenth-century Paris. Ideally, such projects would integrate the remarkable immersive and haptic achievements of historical aural augmented reality projects such as the Virtual Paul's Cross Project with the historicization of sound found in "The Roaring Twenties."[21] Such projects would require teams of specialists from across disciplines that could cooperate toward sonified replication and reenactment. Shawn Graham and his cohort note that the problem with aural augmented reality projects remains how to bring visitors to hear in a historically situated way.[22] Though ruptures between contemporary and historical understanding can elicit productive cognitive dissonances, "Organs of the Soul" demonstrates that the kind of information necessary to attempt historically situated listening lurks in past forms of recording technology including archival documents, musical notation, and surviving objects and architectural structures. The question becomes: How can scholars sonify this information and present it in a format that welcomes visitors into the reenactment of historical acoustemologies?

The Projet Bretez, an interinstitutional team of scholars and engineers across France led by musicologist Mylène Pardoen, provides the replica that complements my quest for eighteenth-century Parisian audition presented in "Organs of the Soul."[23] A historical aural augmented reality project inspired by the experience of video games, Bretez reconstitutes, in great detail, a sound walk through the Châtelet area of 1730s Paris, and eventually the project will be installed for public view.[24] The team also hopes to develop the installation into an immersion room, make it accessible through virtual reality goggles, and create an application for tablets and smartphones. Project leader Pardoen identifies two goals of Projet Bretez: to recuperate the material dimension of sounds from the past and to create an augmented reality of quotidian sound.[25] At first, Bretez visitors hear only their own breath and footsteps, as they peruse a map of 1730s Paris, then, as they enter into the streets, their ears are filled with sounds of crowd commotion, tavern music, birds cawing, and water dripping, while they walk past exacting replicas of buildings that once stood on and around the bridges that cross the Seine. In the spirit of augmented reality, visitors are supposed to experience 1730s Paris not through an avatar but as themselves. Sounds for the project were recuperated from earwitness testimonies, maps, and drawings, among other historical sources, and replicated through the use of period objects and machines. Through careful acoustic modeling work, engineers are currently developing reverberation and echoes true to the architectural spaces in which the visitor is immersed.

Historical aural augmented reality projects aim to create productive dissonances between the past and the present by presenting familiar experiences that visitors can grasp while also pushing against their modern assumptions.[26] The common experience of bustling city street life serves as modern listeners' entry point into this eighteenth-century simulation. As Pardoen explains, even today we are familiar with the density and collective experience of city life, and it is this common point of reference between past and present that should facilitate interaction with the project.[27] Despite the sound of crowds heard throughout the walk, Bretez does not visualize eighteenth-century people for both practical and intellectual reasons. The creation of numerous individuals to inhabit the project space would require a significant logistical undertaking, and a crowded virtual landscape could slow communication between servers and devices when the project ultimately becomes a tablet and smartphone application. It would also be impossible to create the physical sensation of a crowd to corroborate the sonic and visual representation. To justify this decision, Pardoen notes that people tend to walk in a city with their eyes lowered, hearing their environment while not particularly regarding it.[28] Thus, an attempt to block out fellow city dwellers should function as one of the commonalities between eighteenth-century and modern urban walks. Most interesting, though, Pardoen explains that if people were included in the virtual landscape, careful consideration would have to be given to how the eighteenth-century French language sounded, and the team feels that too many questions linger on this issue to confidently include discernable speech.[29]

The Bretez team's insistence on fidelity to past sounds and its distrust of auditioning subjects resonates with the parameters of the Virtual Paul's Cross Project. While recuperating historical sounds, the Bretez team attempted to distinguish objective descriptions of sound provided by eighteenth-century earwitness accounts from the sentiments or interpretations expressed by those auditioning subjects about their sonic experiences. In addition to invisible people, scholars of eighteenth-century France will note a salient lack of bells in the Paris sound walk. On a practical level, because the project is currently meant as a prototype for museums, the team avoided the incorporation of sounds above a certain decibel level in consideration of museum employees who would listen to the project on repeat. From an academic perspective, though, Pardoen felt that the inclusion of bells would require an explanation of the language of bells, an issue she believes would concern scholars more than museum visitors and the general public.[30] In essence, bells might alienate modern listeners from the commonality achieved through the concept of a city walk. If the Bretez team introduced bells into their replication, they would be forced to confront auditioning subjects, who offer the key to understanding this historical sonic marker. A similar concern motivates the two pages in "Organs of the Soul" that address the language of bells and describe both the significance of bells in eighteenth-century French communities and the revolutionary context that resulted in the confiscation of these sonic markers.[31] Pioneering historical work by Alain Corbin has shown that bells delimited time and space in a way that was crucial to everyday French life before the revolution.[32] Ideally, a digital audible history project could present the language of bells not as textual marginalia, as it is presented in "Organs of the Soul," but within the context of a soundscape replication like Projet Bretez. Though Bretez shares my interest in historical audition, its conviction to faithfully excavate past sounds necessitates that emotions attached to sounds be parsed out from the recovery of sensorial experience.

When earwitness subjectivity—and perhaps even mishearing—is siphoned off during the excavation of past sounds, digital audible history projects miss an opportunity to create dialogue between historical auditioning subjects and modern listeners. To return to a previous example, the siren stories would be far less compelling without my own experience elucidating them. A desire to facilitate transhistorical communication motivated the podcasts found on the "music" path of "Organs of the Soul."[33] The podcasts present eighteenth-century Parisian debates about the merits of French and Italian opera from various historical perspectives, including those of composers, men of letters, salon women, and the royal family. The podcasts were the result of a semester-long master's seminar I taught titled Quarrelling about Opera in Eighteenth-Century France. The conception and production of the project were completely in the hands of music graduate students in my course who carefully studied academic sources on the topic. Narrators in the podcasts speak in the present tense, and so the podcasts are a type of historical reenactment presented through a twenty-first-century medium. Music illustrates well how modern listeners assume they understand a sonic experience merely because it exists in both the past and the present, even though modern ears could not possibly hear the political, social, and cultural debates that underpin, for example, eighteenth-century descriptions of Italian music as spicy or French music as refined. Listeners in eighteenth-century Paris, like listeners in any time and place, experienced music within a unique context. As Bretez takes the familiar experience of collective city life as its point of entry, the podcasts employ a familiar medium to present historical arguments in an engaging, haptic form. Podcasts tune modern listeners to current news and debates, and in this case to contemporaneous issues in eighteenth-century Paris. Of course, the podcast reenactments required creative liberties—silly accents to help the listener distinguish between a complicated cast of characters, background noise to create space in the listener's mind, and invented characters to develop a straightforward and entertaining narrative. These liberties, however, do not detract from the careful academic research and debate that produced these podcasts, evinced by the traditional "footnotes" that annotate the podcasts. A modern listener can acquire from the podcasts the sonic knowledge that informed how eighteenth-century auditioning subjects in Paris experienced opera. The podcasts might be critiqued as historical fiction, but even so, they offer a solution to pulling historical acoustemologies from the textual marginalia and into modern sensory experience.

As I write this, I watch a little girl playing in a park sandbox. Her mother holds a tiny sifter, demonstrating how to strain the sand and to search for objects. On the most palpable level of experience, the child is merely playing. Developmentally, though, she is learning to use a tool, to search systematically, and to evaluate objects. The metaphor here for my vision of digital audible history is both fortuitous and striking—both academics and the public should enjoy opportunities to "get dirty" and "hold the sifter" during the digital excavation of historical acoustemologies.[34] Each group will take away different knowledge from the process, of course, but these audiences need not be relegated to separate sandboxes. Admittedly, institutional silos cause some roadblocks to this kind of inclusive scholarship because massive projects like Bretez, for example, require significant funding. A flow chart depicting the transdisciplinary actors and tools that make up Projet Bretez includes scholars from four humanistic disciplines; web developers; information scientists; experts in urban studies, geography, and archaeology; and innumerable digital platforms and providers of technological support. To successfully obtain funding for such complex projects, grant writers must often make a case for the widest possible impact. In these public iterations, academic considerations might be sidelined in the final product. This false dichotomy between the public and the academic also rests at the heart of the authenticity and reenactment issues that I raised previously. The troubling underside of those debates assumes that the public cannot think and that academics cannot play. Digital audible history might breach this barrier and invite the public to engage more critically in the recovery of history and allow academics to immerse themselves in the reenactment of sources. Such a conceptual shift would require funding institutions to reconsider the rigid definition of audiences that is often required by grant applications. Subsequently, universities would need to reconsider the kinds of scholarship that support tenure and promotion cases.

The strength of digital audible history rests in its ability both to foreground sounds recovered from the past and to simulate knowledge carried within, around, and among sounds from a particular time and place. In the twenty-first century we cannot comprehend the word "citizenship" with eighteenth-century minds, but we can work toward an understanding of eighteenth-century conceptions of "citizenship." Just as scholars reconstruct concepts around words before interpreting them historically, concepts around sounds must be reconstructed before we can understand how they were heard. Digital audible history should not only recover and reconstruct sounds, but, more importantly, it should also reanimate historical acoustemologies. Anxieties about inauthenticity and anachronism in digital audible history reveal how traditional academic formats like books and articles also mediate historical material, demanding that scholars confront themselves as a medium, as well. The materiality of digital formats demands reflection upon how writing has also both facilitated and obscured our insight into the past, and how archival research is a contingent practice performed within an institutionalized set of discourses that can never holistically or authoritatively represent historical experience. The challenges of digital audible history reveal the extent to which methodological and theoretical assumptions rest in the very form of scholarship. Therefore, the digital reconfiguration of sonic artifacts, which sometimes performs and reenacts archival materials, should not be considered inauthentic or anachronistic. Rather, it should be understood as an effort to engage past auditioning subjects in the present to create a new archive for the future.[35] One might ask what digital audible history is *for*. Digital audible history both recovers past sounds and reanimates past acoustemologies. This goal requires not only replicas, which imply a distanced, museum-like regard, but also reenactment, which implicates and engages both the scholar and her audiences in confrontations with historical acoustemologies.

NOTES

1 Feld, *Jazz Cosmopolitanism*, 49, and "Acoustemology."
2 Feld, *Jazz Cosmopolitanism*, 131.
3 These methods have particularly flourished in the subfield of archaeology called archaeoacoustics. Researchers have worked toward establishing stricter methodologies; see, for example, Debertolis et al., "Research for an Archaeoacoustics Standard." High-profile projects that have stemmed from the field include a reconstruction of the sound of Stonehenge (Till et al., Sounds of Stonehenge). Archaeoacoustic research has been used to develop aural augmented reality apps, which enhance modern experiences of historical or ancient sites. For example, an iPhone app developed in consultation with Till's team displays what Stonehenge would have looked like as visitors walk around the site, and, through headphones, plays reconstructions of the stones' echoes in various locations. Shawn Graham et al. explore recent work in the field of archaeoacoustics in "Hearing the Past."
4 I specify the term "pre-recording-technology" to denote time periods for which we have no sound recordings such as vinyl records, films, tapes, compact discs, etc., within the archive. I chose this term as opposed to "premechanical reproduction," which could encompass much earlier technologies such as the printing press, barrel-pin plates, and more.
5 In a panel on "Embodying the Past: The Rewards and Risks of Reenactment," convened at the 2014 annual meeting of the American Society for Eighteenth-Century Studies (fittingly held in Williamsburg, Virginia—a Mecca of living history), panelists and audience members engaged in a fruitful dialogue about the anxieties and challenges faced by eighteenth-century scholars who support or participate in reenactment as a means of academic inquiry. The discussion became a type of group therapy in which scholars "came out" as believers or participators in living history—embodying the past in the present.

Schneider's *Performing Remains* astutely reveals the tangible historical work that reenactment, and specifically reenactments of Civil War battles, achieves. She asserts that reenactors "engage in this activity as a way of accessing what they feel the documentary evidence upon which they rely misses—that is, live experience" (10). This emphasis on live experience becomes paramount in time-based art, or historical evidence that is considered ephemeral (for example, sound or music). Schneider concludes that in its desire to preserve the ostensible purity of written archival traces, mimesis becomes debased as a means of accessing the past. Conversely, the performance of the past, of these archival traces, in the present negotiates a new archive for the future that is not solely dependent upon a monomaniacal belief in the written (silent) archive as authoritative.

6 For background on debates that surrounded historically informed performance practice during the 1980s and 90s in the discipline of musicology, see Butt, *Play with History*, 3–52.
7 Brady, *Spiral Way*.
8 Erlmann, *Reason and Resonance*; Johnson, *Listening in Paris*; Nancy, *Listening*; Szendy, *Listen*.
9 Schafer, *Soundscape*, 8–9; see also Rodaway, *Sensuous Geographies*.
10 Corbin, *Village Bells*; M. Smith, *Listening to Nineteenth-Century America*; M. Smith, *Hearing History*; B. Smith, *Acoustic World*; Thompson, *Soundscape of Modernity*; Rath, *How Early America Sounded*; Birdsall, *Nazi Soundscapes*; and Ochoa Gautier, *Aurality*.
11 The distinction between acoustic space and soundscapes is set forth in Mlekuz, "Listening to Landscapes."
12 Virtual Paul's Cross Project. Articles by Wall resulting from the project include "Transforming the Object of our Study," "Recovering Lost Acoustic Spaces," and "Virtual Paul's Cross."
13 I use the term "intellectually rigorous" because the Paul's Cross team applied principles from the London Charter for the Computer-based Visualization of Cultural Heritage to acoustic realizations and modeling, although analogous standards for such acoustic projects do not yet exist. The London charter establishes "internationally recognised principles for the use of computer-based visualisation by researchers, educators and cultural heritage organisations." The full text can be found at www.londoncharter.org (accessed November 28, 2017).
14 Blesser, *Spaces Speak, Are You Listening?*
15 Virtual Paul's Cross Project. "Outcomes." Accessed November 28, 2017. http://vpcp.chass.ncsu.edu/outcomes.
16 Virtual Paul's Cross Project. "The Interactive Sermon." Accessed November 28, 2017. http://vpcp.chass.ncsu.edu/listen-interaction.
17 Thompson, "The Roaring Twenties."
18 Thompson, *Soundscape of Modernity*.
19 See, for example, Johnson, *Listening in Paris*; Nancy, *Listening*; and Szendy, *Listen*.
20 Geoffroy-Schwinden, "Organs of the Soul."
21 See Graham et al., "Hearing the Past."
22 Graham et al., "Hearing the Past."
23 Bretez Site Officiel.
24 A prototype of the project on YouTube can be found through Cailloce, "Écoutez le Paris du XVIIIe siècle."
25 Mylène Pardoen (principal investigator), personal communication with the author, June 21, 2016.
26 Graham et al., "Hearing the Past," describe the concept as "breaks" that focus a participant's attention. For example, when visitors experience the Bretez Project in Châtelet, they will likely be struck by the discordance between

the simulation and their modern experience of Châtelet, and in turn, this should cause visitors to focus more thoughtfully on histories of their current surroundings.

27 Pardoen, personal communication with the author, June 21, 2016.
28 Pardoen, personal communication with the author, June 21, 2016.
29 Pardoen personal communication with the author, June 21, 2016; and Pardoen, email message to author, June 24, 2016.
30 Pardoen, personal communication with the author, June 21, 2016, and Pardoen, email message to author. On the language of bells in eighteenth- and nineteenth-century France, see Corbin, *Village Bells*.
31 Geoffroy-Schwinden, "Organs of the Soul: Sound," 5, 6.
32 Corbin, *Village Bells*; also see note 10 above.
33 Geoffroy-Schwinden, "Organs of the Soul: Music," 2.
34 In an interview with Gita Manaktala, Thompson articulated a similar goal for "The Roaring Twenties." Manaktala, "Aural History on the Web."
35 Schneider, *Performing Remains*; also see note 5 above.

WORKS CITED

Birdsall, Carolyn. *Nazi Soundscapes: Sound, Technology and Urban Space in Germany, 1933–1945*. Amsterdam: Amsterdam University Press, 2012.

Blesser, Barry. *Spaces Speak, Are You Listening? Experiencing Aural Architecture*. Cambridge, MA: MIT Press, 2007.

Brady, Erika. *A Spiral Way: How the Phonograph Changed Ethnography*. Jackson: University Press of Mississippi, 1999.

Bretez Site Officiel. Accessed June 21, 2016. https://sites.google.com/site/louisbretez.

Butt, John. *Play with History: The Historical Approach to Musical Performance*. Cambridge, UK: Cambridge University Press, 2002.

Cailloce, Laure. "Écoutez le Paris du XVIIIe siècle." CNRS *Le Journal*, June 16, 2015. https://lejournal.cnrs.fr/articles/ecoutez-le-paris-du-xviiie-siecle.

Corbin, Alain. *Village Bells: Sound and Meaning in the Nineteenth-Century French Countryside*. New York: Columbia University Press, 1998.

Debertolis, Paolo, Slobodan Mizdrak, and Heikki Savolainen. "The Research for an Archaeoacoustics Standard: The Right Approach to This New Complementary Science for Archaeology." Paper presented at the Second Virtual International Conference on Advanced Research in Scientific Areas (ARSA–2013), Slovakia, December 2–6, 2013. Accessed June 30, 2015. www.sbresearchgroup.eu/index.php/en/articoli-in-inglese/206-the-research-for-an-archaeoacoustics-standard.

Erlmann, Veit. *Reason and Resonance: A History of Modern Aurality*. New York: Zone Books/Cambridge, MA: Distributed by MIT Press, 2010.

Feld, Steven. "Acoustemology." In *Keywords in Sound*, edited by Matt Sakakeeny and David Novak, 12–21. Durham, NC: Duke University Press, 2015.

Feld, Steven. *Jazz Cosmopolitanism in Accra: Five Musical Years in Ghana*. Durham, NC: Duke University Press, 2012.

Geoffroy-Schwinden, Rebecca Dowd. "Organs of the Soul: Sonic Networks in Eighteenth-Century Paris." In *Provoke! Digital Sound Studies*, edited by Mary Caton Lingold, Darren Mueller, and Whitney Trettien. Duke University Franklin Humanities Institute PhD Lab in Digital Knowledge, Soundbox Project, 2014. Accessed November 28, 2017. http://soundboxproject.com/project-organs.html.

Graham, Shawn, et al. "Hearing the Past." *Electric Archaeology*, January 5, 2015. http://electricarchaeology.ca/2015/01/05/hearing-the-past.

Johnson, James H. *Listening in Paris: A Cultural History*. Berkeley: University of California Press, 1996.

Manaktala, Gita. "Aural History on the Web: Reconstructing the Past through Sound." MIT Press blog. Accessed June 30, 2016. https://mitpress.mit.edu/blog/aural-history-web-reconstructing-past-through-sound-0.

Mlekuz, Dimitrij. "Listening to Landscapes: Modeling Past Soundscapes in GIS." *Internet Archaeology* 16 (2004): n.p. http://dx.doi.org/10.11141/ia.16.6.

Nancy, Jean-Luc. *Listening*. New York: Fordham University Press, 2007.

Ochoa Gautier, Ana Maria. *Aurality: Listening and Knowledge in Nineteenth-Century Colombia*. Durham, NC: Duke University Press, 2014.

Rath, Richard Cullen. *How Early America Sounded*. Ithaca, NY: Cornell University Press, 2003.

Rodaway, Paul. *Sensuous Geographies: Body, Sense, Place*. London: Routledge, 1994.

Sakakeeny, Matt, and David Novak, eds. *Keywords in Sound*. Durham, NC: Duke University Press, 2015.

Schafer, R. Murray. *The Soundscape: Our Sonic Environment and the Tuning of Our World*. Rochester, VT: Destiny Books, 1993.

Schneider, Rebecca. *Performing Remains: Art and War in Times of Theatrical Reenactment*. London: Routledge, 2011.

Smith, Bruce R. *The Acoustic World of Early Modern England: Attending to the O-Factor*. Chicago: University of Chicago Press, 1999.

Smith, Mark M. *Listening to Nineteenth-Century America*. Chapel Hill: University of North Carolina Press, 2001.

Smith, Mark M., ed. *Hearing History: A Reader*. Athens: University of Georgia Press, 2004.

Szendy, Peter. *Listen: A History of Our Ears*. New York: Fordham University Press, 2008.

Thompson, Emily. "The Roaring Twenties: An Interactive Exploration of the Historical Soundscape of New York City." *Vectors: Journal of Culture and Technology* in a Dynamic Vernacular (Fall 2013): n.p. Accessed November 28, 2017. http://vectorsdev.usc.edu/NYCsound/777b.html.

Thompson, Emily. *The Soundscape of Modernity: Architectural Acoustics and the Culture of Listening in America, 1900–1933*. Cambridge, MA: MIT Press, 2002.

Till, Rupert, et al. Sounds of Stonehenge. Accessed June 30, 2015. https://soundsofstonehenge.wordpress.com.

Virtual Paul's Cross Project. "A Digital Re-creation of John Donne's Gunpowder Day Sermon." Raleigh: North Carolina State University. Accessed November 27, 2017. http://vpcp.chass.ncsu.edu.

Wall, John N. "Recovering Lost Acoustic Spaces: St. Paul's Cathedral and Paul's Churchyard in 1622." *Digital Studies/Le Champ Numérique*, Proceedings of the SDH-SEMI 2012 Conference. Accessed November 28, 2017. www.digitalstudies.org/articles/10.16995/dscn.58.

Wall, John N. "Transforming the Object of Our Study: The Early Modern Sermon and the Virtual Paul's Cross Project." *Journal of Digital Humanities* 3, no. 1 (Spring 2014): n.p. http://journalofdigitalhumanities.org/3–1/transforming-the-object-of-our-study-by-john-n-wall.

Wall, John N. "Virtual Paul's Cross: The Experience of Public Preaching after the Reformation." In *Paul's Cross and the Culture of Persuasion in England, 1520–1640*, edited by Torrance Kirby and P. G. Stanwood, 61–92. Leiden: Brill, 2014.

SOUND PRACTICES FOR DIGITAL HUMANITIES

STEPH CERASO

> In much of our experience our different senses do not unite to tell a common and enlarged story.
> — JOHN DEWEY, *Art as Experience*

The coaster steadily clicks and clacks until it gets to the peak of the hill, slowing until there are just a few punctuated beats—then comes the whooshing rush of the stomach-sinking drop. Riding a rollercoaster is an immersive, holistic sonic experience. Wooden rollercoasters elicit an especially unique affective response that cannot be replicated in smoother, faster, and much quieter steel speed coasters. The distinct thunderous roar produced by the wooden tracks is experienced not only through the ears but through the entire body. The feeling of the rattling, clattering sound is what propels and intensifies the journey for riders; their bodily experiences are inextricably linked to the jarring sounds of the coaster.

While not every sonic encounter is as exhilarating as a rollercoaster ride, it is not unusual to experience sound as a sensorially invigorating event. Standing near the stage at a concert, playing an instrument, or simply driving one's car when a Mack truck zooms by are multisensory, as opposed to merely auditory, sonic experiences. Though listening is almost always associated with the ears, these examples make clear that the experience of sound is not limited to a single sense. Indeed, the convergence of sight, sound, and touch (and sometimes smell and taste) is in part what makes sonic interactions so engrossing and compelling.

Yet when sound is incorporated into digital environments, the multisensory potential of sound tends to be dismissed or forgotten. Specifically, in most digital scholarship sound is treated as a *semiotic resource* rather than an *experience*. Sound files that are embedded into websites, blogs, digital archives, and other audio projects are used to enhance or exemplify the content of an accompanying textual narrative, to serve as narratives in and of themselves (e.g., podcasts, interviews), or to enrich visual media (e.g., digital maps, video soundtracks, and voiceovers). Other than the fact that sound can be heard—for those of us with working ears—there is not much difference between the typical ways that sound is incorporated into digital scholarship and more traditional alphabetic forms of scholarship. Like text, sound is presented as information that is ripe for interpretation and analysis. The dissemination of meaningful sonic information in digital scholarship takes precedence over users' embodied experiences—the ways in which users physically interact with and are affected by sound at the level of the senses.

There is certainly scholarship that calls for or works best in more traditional audio formats. However, to approach sonic scholarship exclusively as another site of meaning making is to ignore both the distinct affordances of composing in digital contexts and the fact that living, sensing, nerve-filled human bodies—not just ears and brains—interact with it.[1] As the editors note in the introduction to this collection, "Dealing with sound means dealing with the lived experience of people." Alongside digital projects that resemble familiar textual scholarship, how might the *lived experiences* of listeners play a more salient role in the production of digital sound studies work? How might we account for and learn from the sonic experiences of *all bodies*—including deaf or hard-of-hearing individuals—and how might such bodily experiences inform digital design?

This essay proposes several "sound practices" that are intended to help scholars account for fully embodied kinds of sensory engagement; these practices amplify the ecological relationship between sound, bodies, and environments. The term "sound practices" refers to more than the literal sounding of digital scholarship. The definition of sound as "exhibiting or based on thorough knowledge and experience" resonates with these practices as well.[2] Drawing from and extending this definition, sound practices are intended to encourage scholars to produce work that is grounded in a thorough knowledge of the ideas they are exploring *and* a thorough knowledge of the diverse ways that bodies experience sound. That is, sound practices imply a thoughtful consideration of one's own and others' embodied listening experiences during the processes of creating, designing, publishing, and interacting with digital sound scholarship.

As evidenced by the range of dynamic work in this collection, we are still in the early, noisy stages of figuring out what "counts" as digital sound studies; it is an area of digital humanities that is being invented as it grows. With the potential for invention and growth in mind, this essay adopts a generous, roomy conception of digital sound studies that includes the type of small-scale sonic experiments featured in the digital counterpart to this book. While the sound practices identified will ideally prove useful to digital scholars writ large, they are aimed at those interested in using sound to immerse listeners in sensory-rich experiences. Such creative-critical projects can energize and broaden the scope of digital sound studies (and digital humanities) by emboldening scholars to take more imaginative, playful, and inclusive approaches to sonic scholarship.

Sound practices ask scholars to rethink and work around the constraints of digital composition—from two-dimensional screen space to the limited audio capabilities of digital devices—to produce more holistic sonic experiences. What follows, then, is an exploration of the various possibilities for creating digital sonic interactions that go beyond exclusively ear-centric modes of listening; for producing digital sonic experiences that are more similar to the kinds of intense, affectively powerful experiences of sound in physical environments; and for designing heightened, flexible, and immersive sensory experiences in digital contexts. In the spirit of this collection, the sound practices outlined below serve as provocations for initiating more substantive conversations about the multisensory possibilities of digital sound studies scholarship.

Sound Practice #1

Consider how different bodies with a range of sensory capacities and diverse needs might interact with sound-based digital projects.

Thinking about how audiences will intellectually respond to and make use of sound in digital scholarship is a standard practice. I would argue, however, that there needs to be more emphasis placed on accounting for how different kinds of bodies might interact with and have access to sound in digital spaces. Rather than assuming that all bodies are uniform—that all listeners listen in the exact same way—composers of sonic scholarship need to acknowledge and plan for an audience that consists of a diverse range of bodies with various sensory capacities and learning needs.

As a starting point for creating more inclusive digital experiences, scholars working with sound could benefit greatly from having more explicit conversations with disability studies scholars (and vice versa).[3] In recent years sound studies scholars such as Mara Mills and Gerard Goggin have begun to explore disability in relation to media history and technological innovation.[4] While issues of disability and access have been spurring lively discussions in textual sound studies scholarship, they have been largely ignored in the actual production of sonic scholarship. As a result much digital audio work is still being created for an "ideal" listener, thus excluding a broad swath of the population with disabilities and learning needs—most notably people who are deaf or hard-of-hearing—from interacting with and contributing to this work.

Increasing access should mean more than making sonic material available and presenting it in ways that will be useful for scholars from different disciplinary backgrounds. It is also critical to provide users with multiple modes and pathways to engage with and understand sonic scholarship; *flexibility* must become a key part of the design process. In other words, increasing accessibility will require scholars to practice universal design. As Jay Dolmage and Cynthia Lewiecki-Wilson write, universal design is a concept that "holds that one should design spaces and learning environments for the broadest possible access."[5] Adopting universal design as a fundamental practice is a necessary and critical step toward the creation of more inclusive sound-based work in digital environments.[6]

Universal design played a central role in the development of my own digital sound experiment for *Provoke!* My project, "A Tale of Two Soundscapes," examines the relationship between sound and embodied experience in two strikingly different sonic environments: a small town near the Smoky Mountains of North Carolina and the city of Pittsburgh, Pennsylvania. The approach to listening that I enact in "A Tale of Two Soundscapes" offers an alternative to strictly ear-centric modes of listening and amplifies sound as a multisensory experience. The creation of this audiovisual narrative, as well as the extensive research that emerged from it, heightened my awareness of the unique ways that different bodies engage with and make sense of sound in a range of environments. In the digital context I was designing for, one of my goals was to provide multiple ways for users with various listening capacities and preferences to interact with and understand my project. Thus, users who may not be able to hear the audio can read the transcript, which makes available both the written language for my voiced script and descriptions of the nondiscursive sounds that occur in conjunction with my voice. In addition, I included a video of still images that helps to visually contextualize the two contrasting soundscapes that are being explored, and the script provides time markers for users who want to compare what is happening in the text with what is happening in the video.

Significantly, users have the option to interact with these different media in whichever way(s) best suits their needs and purposes. By offering users various choices for accessing the same material, I tried to follow the lead of disability studies scholars like Stephanie Kerschbaum, who writes, "Those who design and produce multimodal texts and environments need to incorporate redundancy across multiple channels in order to make digital texts more—not less—flexible, and they should enable customization and manipulation of these texts."[7] While developing "A Tale of Two Soundscapes," I became acutely aware of the ways that redundant design can effectively facilitate multiple pathways for interaction. As a result, this piece was designed to give users the option to listen to the audio only, or to listen to the audio while following along with the transcript, or to read the transcript on its own or while interacting with the video, or to listen to the audio and watch the video simultaneously. I also included longer clips of the isolated soundscape tracks for listeners who want a chance to focus on individual field recordings without the distraction of a voiceover. These longer clips provide a point of comparison for the ways in which the field recordings were manipulated and edited in the main audio narrative, calling attention to the mediation inherent in the creation of sound-based digital work. As I found out, redundant design serves a dual purpose: it gives users multiple ways to engage, and in doing so it makes other forms of intellectual work possible. In this case, it makes the process of composing transparent (and therefore available for examination) by revealing how I did and did not alter original field recordings. Accessible design, then, need not be approached in a strictly practical way; it has both utilitarian and intellectual functions.

Though I tried to make my project accessible to a broad audience, I do not mean to suggest that it is ideal for every user in every situation. That would be impossible. As the authors of "Multimodality in Motion" remind us, "Universal design is a process, a means rather than an end. There's no such thing as a universally designed text. There's no such thing as a text that meets everyone's needs. . . . But to say that no text will be universally accessible is not a justification for failing to consider what audiences are invited into and imagined as part of a text."[8] Universal design is something that everyone can strive for and work toward. Choosing to think seriously about who might be listening to and interacting with our work will open up new possibilities for who is "invited into and imagined as being a part of" the digital sound studies community.[9]

At the same time, it is important to recognize that designing projects for an *abstract* broadest possible audience is not enough. It is also essential to consider how individual users actually interact with published work. One of the advantages of publishing digital scholarship is that it does not have to remain static and fixed like most print publications. The fact that digital work can be changed and revised gives authors a chance to get feedback from individual users and continue to tweak their work based on reported suggestions or accessibility issues.[10] Scholars can encourage such feedback by providing statements of access on their main project pages, including contact information, so that users with questions or concerns can reach them directly.[11]

As Dolmage convincingly argues, accounting for both universal design and usability, or how people are able to interact with digital scholarship (or not), can result in productive conversations that get projects closer to achieving the broadest possible access.[12] Rather than only making "corrections" to digital audio work because of accessibility complaints—what is referred to as "retrofitting" in the disability studies community—relying on both universal design and usability is a way for authors to produce scholarship that is widely accessible from the start, and to collaborate with users via discussing and discovering new ways of inclusion.[13]

Increasing access to sound scholarship is necessary first and foremost because all listeners—regardless of sensory capacities and bodily needs—deserve the opportunity to participate in and contribute to this exciting and steadily growing area of digital humanities. Further, the participation and contributions that result from increased access could expand and augment digital sound studies in important ways. I could envision, for instance, digital audio projects by deaf scholars that enact their individual sonic experiences, thus contributing to understandings of listening as a body-specific, multisensory practice; or perhaps digital work that focuses on how various kinds of embodied experiences influence the ways in which individuals respond to and make sense of sound in different contexts. Indeed, though I have chosen to focus on disability in this discussion, it is equally important to represent and perform embodied experiences of race, gender, class, age, ethnicity, and sexual orientation in digital scholarship. Such bodily experiences have a profound effect on how people engage with the sonic world. If a more diverse range of bodies and bodily practices was welcomed into and encouraged to take part in the conversation, imagine what insights and boundary-pushing projects might emerge.[14] Cultivating a more inclusive digital sound studies community by devoting substantial attention to embodied experiences will lead to a richer, more capacious intellectual and creative space for digital scholarship.

Sound Practice #2

Take fuller advantage of the spatial and aesthetic features of digital sound projects to create more immersive user experiences.

In contrast to the immersive experience of sound in three-dimensional spaces, it is easy to forget that sound in digital spaces is located in an environment at all. If they can see, listeners engage with sound while looking at flat, two-dimensional images on a screen. If they can hear, they listen through minuscule speakers or tiny earbuds that diminish the effects of sound. Though sonic composition for digital environments has its limitations, it seems to me that scholars can enliven the *experience* of their sonic work by taking fuller advantage of spatial and aesthetic affordances in digital spaces. That is, in addition to treating sound as an object that is the analytical focal point of digital sound studies scholarship, we might also use sound as a way to create more dynamic digital environments—digital spaces that bodies navigate and experience via multiple senses.

One way to create more immersive sonic experiences for users is to learn to think more like acoustic designers. Acoustic designers (sometimes called acoustic engineers) are sound professionals who design, change, and/or enhance the acoustical environment of particular spaces—from restaurants to concert halls to parks. Though acoustic design is a complex interdisciplinary field, here I want to amplify a basic acoustic design principle that I find relevant for digital sound studies: *acoustic designers treat sound as an element that is connected to and influenced by a larger aesthetic and spatial network*.[15]

Consider, for example, the acoustic design of the lobby of an office building.[16] The lobbies of buildings are places where socializing is expected, and thus they are designed to be sonically lively places. To add some extra noise and life into the space, acoustic designers would design or manipulate the spatial and aesthetic features of lobbies—via ample open space, high curved ceilings, hard surfaces like marble or concrete—to produce a reverberant environment, or a space where sound persists after the original sound is produced. Reverberation makes it seem as if there is more sound filling a space than actually exists, giving the space a warm, energetic atmosphere that makes people feel like it is appropriate to talk loudly and be more social. However, since the rest of the building is dedicated to traditional office space where acoustics need to enable productive (i.e., less disruptive) working conditions, acoustic designers would design a quieter, deader acoustical environment to cue people to be less animated and social as they move through the building. In other words, the acoustics of the space would need to be designed to signal people to adjust their behavior accordingly: the rooms would be smaller and box-like to prevent reverberation, the hallways and office walls would be built with more insulation or sound-absorbing materials, and so on. Good acoustic designers are always conscious that the ways people experience and respond to sound in an environment are inextricably connected to the aesthetic and spatial features of the design.[17]

While digital spaces are significantly different from three-dimensional spaces like the lobby of a building, being more cognizant of sound as a design element that is connected to and shaped by other features of an environment can help scholars produce more cohesive, immersive projects. Taking advantage of the spatial and aesthetic affordances of digital audio involves considering questions such as: How do I want listeners to move through and experience my project? How might I make the various digital spaces of my project more sonically distinct from one another? How does the experience I created enact the themes or arguments or stories I want to present? How do the aesthetic features (colors, textures, layout) influence the ways that listeners might experience sound? How can I enable nonhearing individuals to experience a sonic project, and in turn, how might addressing issues of access lead to a better design in general? In sum, approaching digital work like an acoustic designer requires thinking about sonic scholarship as a *holistic experience* for users.

Sharon Daniel's digital project "Public Secrets" serves as an excellent model of creative-critical sound scholarship that is designed with the holistic experiences of users in mind. In "Public Secrets," Daniel takes users along with her into the sprawling prison-industrial complex in central California to hear the testimonies of women prisoners. There are many interesting features of the design, which masterfully integrates sound, text, visual elements, and movement. What I find most striking, though, is the way that Daniel uses sound to draw (hearing) listeners into the experience.[18] In the opening sequence she verbally describes the scene of the prison. Her vocal track is layered with heavy music—a sorrowful, repetitive melody punctured by snare drums—as well as the ambient soundscape of the prison itself. The layered sounds immediately position listeners within the environment of the prison while evoking the tone or feeling of the space.

Once users officially enter the project, they can choose different theme-based pathways to navigate through it. Clicking on these themes triggers more startling sounds: the creaking, locking, and slamming sounds of a heavy iron door. These sounds work to incorporate users into the prison experience. By making users "occupy" the same sonic space as the prisoners, Daniel is blurring the line between inside and outside. Other design elements echo and intensify this blurring. The primarily black-and-white color scheme reflects the drastic differences between inside and outside (and is perhaps meant to conjure up other binaries: good and bad, right and wrong, etc.). However, the algorithmic structure of the project causes the black-and-white spaces of the screen to constantly shift depending on where users click, thus enacting the idea that things are not as clear-cut—as black and white—as they may appear. Sound, color, layout, movement, space. These integrated features of the design all serve to drive home Daniel's main point: that the prison-industrial complex affects all of us, not just the lives of those women on the inside, whose hidden, incarcerated bodies are afflicted with racism, sexism, poverty, abuse, and addiction. We are all implicated in this networked system despite the boundaries we try to create between "us" and "them."

Much more could be said about the content and political implications of Daniel's project. For the purposes of this discussion, though, I want to underscore that the use of nonverbal sound in "Public Secrets" is so effective because it is thoroughly integrated with various aesthetic and spatial features of the design. The sound is not employed as an isolated part of the project but as a salient component of its sensory and thematic experience as a whole. Daniel's consideration of how bodies move through and participate in a space via multiple senses and modes is the key to creating an affectively powerful and thought-provoking experience for users—an experience that could not be accomplished through a more linear (or traditionally academic) version of her work.

By calling attention to "Public Secrets" as an example of what an acoustic design approach might look like in a digital environment, I do not mean to suggest that every digital sound project needs to be or even can be an immersive experience. Clearly, the design one chooses would depend on the purposes and goals of the scholarship. It is also important to recognize that "Public Secrets" is a large-scale undertaking that was made possible through generous institutional funding and the support of *Vectors*, an innovative digital publishing platform. Though not every scholar will have access to such resources, I think that there is still a lot of room in both large- and small-scale digital sound studies work for experimenting; for designing more holistic experiences for users (as opposed to presenting sonic data or information)—for treating sound as an element that is connected to and influenced by the other features of a design. Work like Daniel's has only begun to tap into the possibilities for producing distinctive, compelling digital sound environments. My hope is that her example will inspire more scholars to discover and create sensory-rich sonic experiences in their own projects, regardless of scope and scale.

Sound Practice #3

Explore and experiment with the physical effects of sound in digital contexts.

Digital work regularly takes advantage of the audible and visual possibilities of sound. The simple act of being able to incorporate audible files into digital environments is what caused the initial wave of enthusiasm for sonic forms of scholarship. In recent years, this scholarship has been evolving and extending in more synesthetic ways. For example, there have been an increasing number of sonification projects, such as Listen to Wikipedia and BitListen, that give sound to previously nonsonic information. Additionally, sound and music visualization projects—encouraged by free applications like Sonic Visualiser—are becoming more common in scholarship across the disciplines. However, the physical effects of sound, or the experience of sound as a form of touch, remains a largely uncharted area. This is not especially surprising since the experience of sound is etiolated in digital contexts. Listeners cannot feel the sounds they listen to on computers or phones like they can when they are standing in front of massive speakers at a club. Most digital audio formats and the technologies used to engage with them are not able to re-create these kinds of felt sonic experiences.

And yet, because the physical experience of sound is a significant part of how humans engage with and understand sound, it seems to me like an area that is worthy of sustained inquiry and experimentation. While scholars such as Steve Goodman, Shelley Trower, and Michele Friedner and Stefan Helmreich have written thoughtfully about the physical, vibratory experience of sound in various contexts, I wonder how the physical effects of sound might be *performed* in digital environments.[19] How might digital sound studies scholarship explore and possibly re-create the tactile experience of sound? What would scholarship look, sound, and *feel* like if more attention was paid to sound as a physical event?

Current trends in audio technologies that celebrate vibration as a novel feature of listening experiences may be a productive starting point for investigating the role of touch in digital sonic work. Skullcandy Crusher headphones (advertised as "#bassyoucanfeel") enable listeners to feel the low frequency sounds of bass via vibration. As stated on the Skullcandy website, "Our designers wanted to fix the problem of a single sensory experience with conventional headphones. Combining audio with tactile senses creates a more realistic and immersive environment."[20] Wearable technologies like the 3rd Space gaming vest also use tactile feedback to heighten the experience of sound in video games.[21] Incorporating technologies like these into the design of future digital sound work—or at least presenting them as an option ("This scholarship works best with technology X")—could help introduce tactile possibilities that allow for more fully embodied modes of engagement.

Assistive technologies offer further opportunities for experiencing sound as a form of touch. Psychology professor Frank Russo and his research team recently invented a chair that is intended to enhance musical experiences for deaf and hard-of-hearing audiences. The "emoti-chair," Russo explains, is able to "separate out the frequencies and present them to different parts of the body. We'll take the high frequencies and we'll present them to the upper part of the back. We'll take the lower frequencies in the music signal and we'll present them to the lower part of your back."[22] Rather than simply re-creating a general feeling of vibration, the chair offers a more precise experience of music by pinpointing where certain frequencies resonate in the body. The emoti-chair is a great example of how assistive technologies that were designed for people with disabilities could enrich human experience more broadly. As Graham Pullin points out, specialized products that are created because of "issues around disability [can] catalyze new design thinking and influence a broader design culture in return."[23] Digital sound projects that examine and play with the bodily locations of felt frequencies via technologies like the emoti-chair might facilitate entirely new ways of interacting with digital sound scholarship for everyone—not just people with disabilities.[24] Indeed, as I have stressed throughout this essay, addressing issues of accessibility often results in designs that are broadly beneficial to users as opposed to directed only toward a specific group of users.

Designing projects that involve supplemental technologies will of course raise issues of cost and access. To make tactile experiences a more prominent feature of digital audio work, scholars will need to continue to discuss and troubleshoot the technical and conceptual challenges of creating scholarship with and for these kinds of vibratory audio technologies. However, such projects do not necessarily have to be costly, large-scale endeavors. I could imagine work that takes advantage of the vibratory features of ordinary consumer products like smartphones. A phone application, perhaps, that provides vibratory feedback in relation to an environment's noise level might be an interesting digital tool for making individuals more aware of their embodied experience of sound in different spaces. If an environment has particularly low decibel levels, the phone would automatically buzz intermittently; in environments with high decibel levels, the phone would vibrate more frequently. Tactile feedback would call users' attention to their own physical experiences of sound in a space (something that people often shut out or ignore), thus alerting them to record and geo-tag the decibel information through the app. This hypothetical vibration-based app would enable users to construct a digital map of place-based bodily experiences of sound in their communities, thus helping others to find or avoid the sonic spaces that best suit their needs or preferences.

Of course, the fact that existing technologies present scholars with tactile possibilities does not mean that these technologies should be universally adopted. ("I want to make users feel this sound because I can.") Scholars need to think seriously about how tactile information or force-feedback mechanisms would enhance their work—about what the ability to feel sound in digital scholarship would allow listeners to do or understand that would not be possible (or as effective) using only text, sound, and/or visual elements. That is, digital scholars should consider the distinct affordances of making sound available as a tactile experience for users. While the kinds of technologies mentioned above have limitations and may not be useful for every project, at the very least they have the potential to open up a productive area of inquiry for exploring touch/tactility in digital sound studies.

En*liven*ing Digital Sound Studies

The sound practices I have outlined in this chapter are intended to invigorate digital sound studies scholarship by accounting for the *lived*, multisensory experiences of a broad audience. Adopting and expanding on these practices can result in more engaging, flexible, and affectively powerful sonic compositions and digital tools. To make an impact, however, sound practices cannot merely be taken up by individual scholars. Just as importantly, editors of digital journals who publish sonic scholarship and the institutions that fund such work must be willing to accept and accommodate experimental, sensory-rich, and widely accessible digital sound studies projects. In other words, implementing sound practices is going to require collaboration. No single individual has access to all of the technical skills, knowledge, resources, technologies, and/or bodily experiences that are needed for the kinds of sonic work I have proposed. Thus, as in most digital humanities endeavors, it will be necessary to collaborate to find the right combination of people to turn ideas into reality. As I see it, the challenge of infusing digital sound studies with more experience-based, body-conscious scholarship will be to organize networks of diverse bodies with a range of different needs, capacities, cultural identities, skill sets, disciplinary backgrounds, and professional positions. Such networks will bring us closer to a more inclusive, creatively thriving digital sound studies community—a community that I hope will make enough noise to be seen, heard, and felt in digital humanities.

NOTES

1. Scholarship that explores the senses as integrated rather than separating and/or privileging individual senses has been gaining momentum in recent years, particularly in anthropology and digital media theory. For an excellent overview of this work, see Porcello et al., "Reorganization of the Sensory World."
2. *Merriam-Webster Online*, s.v. "Sound (adj.)," definition 3b, accessed November 29, 2017, www.merriam-webster.com/dictionary/sound.
3. The call to pay more attention to issues of disability and access has been sounded in the larger digital humanities community as well. George H. Williams writes, "It is imperative that digital humanities work takes into account the important insights of disability studies in the humanities, an interdis-

ciplinary field that considers disability 'not so much a property of bodies as a product of cultural rules about what bodies should be or do.'" "Disability, Universal Design," 202.
4. Mills, "Deaf Jam," and "Hearing Aids"; Goggin, "Cellular Disability."
5. Dolmage and Lewiecki-Wilson, "Refiguring Rhetorica," 26.
6. For additional information about the origins of universal design and why it is vital for digital humanities more broadly, see Williams, "Disability, Universal Design."
7. Kerschbaum, "Modality."
8. Yergeau et al., "Multimodality in Motion."
9. The Web Accessibility Initiative of the World Wide Web Consortium website contains guidelines and instructions, as well as links to resources about accessibility and design.
10. Of course, the ability to change and revise digital projects also depends on who is hosting the project, what kind of relationship the host has to the author, and what types of labor people are willing to put into the continuation of a project. Asking the editor or host of one's project about issues of accessibility and possible changes is a good practice, particularly in the early stages of design.
11. I have provided a basic statement of access on the main page of "A Tale of Two Soundscapes" and would welcome feedback. For a brief and helpful explanation of how to write accessibility statements, see Watson, "How to Write an Accessibility Statement."
12. Dolmage, "Evolving Pedagogy," and "Disability, Usability."
13. For more on retrofitting, see Yergeau et al., "Multimodality in Motion."
14. My emphasis on more diversity in digital sound studies scholarship echoes similar calls by a number of digital humanities scholars and organizations. For instance, the position statement created at THATCamp SoCal reads: "We recognize that a wide diversity of people is necessary to make digital humanities function. As such, digital humanities must take active strides to include all the areas of study that comprise the humanities and must strive to include participants of diverse age, generation, skill, race, ethnicity, sexuality, ability, nationality, culture, discipline, areas of interest. Without open participation and broad outreach, the digital humanities movement limits its capacity for critical engagement" (PhDeviate et al., "Towards an Open Digital Humanities"). I see accessible design and an attention to usability at the level of the body as key to achieving more open participation.
15. Thompson's *Soundscape of Modernity* and Blesser and Salter's *Spaces Speak* provide a wealth of information on acoustic design, sound and architecture, and acoustical technologies.
16. This example is based on information from interviews I conducted with professional acoustic designers while doing research for my current book project. For more information on my forthcoming book, visit www.stephceraso.com (accessed November 29, 2017).

17. Sterne's "Sounds like the Mall of America" presents a fascinating, in-depth example of how acoustic environments are designed strategically to persuade people to behave in particular ways.
18. In terms of accessibility, "Public Secrets" provides written transcripts of the prisoners' testimonies. However, one of the limitations of this project is that it does not include captions for nondiscursive sound. Adding textual information for ambient sounds as they occur would further expand access to deaf and hard-of-hearing audiences.
19. Goodman, *Sonic Warfare*; Friedner and Helmreich, "When Deaf Studies"; Trower, *Senses of Vibration*.
20. Skullcandy Crusher: Inspiration behind Bass You Can Feel, March 29, 2013. www.skullcandy.com/blog/2013/03/29/crusher-inspiration-behind-bass-you-can-feel.
21. TN Games, 3rd Space Vest (accessed April 13, 2014, http://tngames.com/products).
22. Mahoney, "Sound (and Sight and Feel)."
23. Pullin, *Design Meets Disability*, xiii.
24. The emoti-chair, for example, has the potential to improve products like the BoomChair, which features "interactive vibration motors" that heighten the experience of sound in video games, music, and movies. The experience of vibration in the BoomChair does not yet provide a location-specific and precise vibratory experience and could thus benefit from the design and technology used in emoti-chairs. BoomChair Official Site (accessed April 12, 2014, www.boomchair.com).

WORKS CITED

Blesser, Barry, and Linda-Ruth Salter. *Spaces Speak, Are You Listening? Experiencing Aural Architecture*. Cambridge, MA: MIT Press, 2007.

Cannam, Chris, Christian Landon, and Mark Sandler. "Sonic Visualiser: An Open Source Application for Viewing, Analysing, and Annotating Music Audio Files." Proceedings of the ACM Multimedia International Conference, 2010. Accessed November 29, 2017. www.sonicvisualiser.org.

Daniel, Sharon. "Public Secrets." *Vectors: Journal of Culture and Technology in a Dynamic Vernacular* 2, no. 2 (2007): n.p. http://vectors.usc.edu/projects/index.php?project=57.

Dewey, John. *Art as Experience*. 1934; reprinted, New York: Perigee, 2005.

Dolmage, Jay. "Disability, Usability, Universal Design." In *Rhetorically Rethinking Usability*, edited by Susan Miller Cochran and Rochelle L. Rodrigo, 167–90. Cresskill: Hampton Press, 2009.

Dolmage, Jay. "Evolving Pedagogy." *Disability Studies Quarterly* 25, no. 4 (Fall 2005): n.p. http://dsq-sds.org/article/view/627/804.

Dolmage, Jay, and Cynthia Lewiecki-Wilson. "Refiguring Rhetorica: Linking Feminist Rhetoric and Disability Studies." In *Rhetorica in Motion: Feminist Methods and Methodologies*, edited by Eileen E. Schell and K. J. Rawson, 23–38. Pittsburgh: University of Pittsburgh Press, 2010.

Friedner, Michele, and Stefan Helmreich. "When Deaf Studies Meets Sound Studies." *Senses and Society* 7, no. 1 (2012): 72–86.

Goggin, Gerard. "Cellular Disability: Consumption, Design and Access." In *The Sound Studies Reader*, edited by Jonathan Sterne, 372–87. London: Routledge, 2012.

Goodman, Steve. *Sonic Warfare: Sound, Affect, and the Ecology of Fear.* Cambridge, MA: MIT Press, 2010.

Hashemi, Mahmoud, and Stephen LaPorte. Listen to Wikipedia. Accessed April 11, 2014. http://listen.hatnote.com.

Kerschbaum, Stephanie. "Modality." In "Multimodality in Motion: Disability and Kairotic Spaces." *Kairos* 18, no. 1 (2013): n.p. http://kairos.technorhetoric.net/18.1/coverweb/yergeau-et-al/pages/mod/index.html.

Laumeister, Maximillian. BitListen. Accessed April 11, 2014. www.bitlisten.com.

Lingold, Mary Caton, Darren Mueller, and Whitney Trettien. "Introduction." In *Provoke! Digital Sound Studies.* Durham, NC: Duke University Press, 2018.

Mahoney, Jill. "The Sound (and Sight and Feel) of Music for the Deaf." *Globe and Mail*, November 9, 2010. www.theglobeandmail.com/technology/science/the-sound-and-sight-and-feel-of-music-for-the-deaf/article1216459.

Mills, Mara. "Deaf Jam: From Inscription to Reproduction to Information." *Social Text* 28, no. 1 (Spring 2010): 35–58.

Mills, Mara. "Hearing Aids and the History of Electronics Miniaturization." In *The Sound Studies Reader*, edited by Jonathan Sterne, 73–78. London: Routledge, 2012.

PhDeviate et al. "Towards an Open Digital Humanities." THATCamp SoCal 2011, January 2011. Accessed April 14, 2014. http://soca12011.thatcamp.org/01/11/opendh.

Porcello, Thomas, Louise Meintjes, Ana Maria Ochoa, and David Samuels. "The Reorganization of the Sensory World." *Annual Review of Anthropology* 39 (2010): 51–66.

Pullin, Graham. *Design Meets Disability.* Cambridge, MA: MIT Press, 2009.

Sterne, Jonathan. "Sounds like the Mall of America: Programmed Music and the Architectonics of Commercial Space." *Ethnomusicology* 41, no. 1 (1997): 22–50.

Thompson, Emily. *The Soundscape of Modernity: Architectural Acoustics and the Culture of Listening in America, 1900–1933.* Cambridge, MA: MIT Press, 2002.

Trower, Shelley. *Senses of Vibration: A History of the Pleasure and Pain of Sound.* New York: Continuum, 2012.

Watson, Léonie. "How to Write an Accessibility Statement." Nomensa.com, February 17, 2009. www.nomensa.com/blog/2009/writing-an-accessibility-statement.

Web Accessibility Initiative. "Web Content Accessibility Guidelines 2.0 (WCAG 2.0, July 2012)." Accessed April 13, 2014. www.w3.org/WAI/intro/wcag.

Williams, George H. "Disability, Universal Design, and the Digital Humanities." In *Debates in the Digital Humanities*, edited by Matthew K. Gold, 202–12. Minneapolis: University of Minnesota Press, 2012.

Yergeau, Melanie, et al. "Multimodality in Motion: Disability and Kairotic Spaces." *Kairos* 18, no. 1 (2013): n.p. http://kairos.technorhetoric.net/18.1/coverweb/yergeau-et-al/index.html.

AFTERWORD

DEMANDS OF DURATION

The Futures of Digital Sound Scholarship

JONATHAN STERNE, WITH MARY CATON LINGOLD,
DARREN MUELLER, AND WHITNEY TRETTIEN

DARREN MUELLER We first started our conversation with you at the early onset of our project. At that time, I remember that our conversation went back and forth quite a bit about the possibilities of technological innovation and what consequences it might have for sound studies. What do you think has changed in the field since then [2012]? Where are digital humanities and sound studies overlapping?

JONATHAN STERNE That's a really long time in the computer industry and it's a really short time in the academy. I don't know that there has been any giant leap forward. It's more like conversations that have been going on for many years have continued. Movements like digital humanities have had a few more years to gain a foothold in the academy. It has become more normal to want to put audiovisual material inside humanities work across all fields, and so people are more comfortable with the idea of using digital technologies in their research and scholarship more generally. The equipment has gotten older, been replaced, been upgraded, and been broken. It's an endless cycle.

One of the things that's really struck me, as I looked at the *Provoke!* website and read the book, is that a lot of the best digital humanities work in sound studies is pretty low-tech. When people want to found a digital humanities lab, they try to get a million-dollar grant or a few hundred thousand dollars, they buy a bunch of computers, get all of these servers, video stuff, high-res scanner, etc. But you don't need to. If you want to start doing digital humanities in an undergrad sound class, almost all of your students, even if they're fairly disadvantaged, have a recording device in their pockets. The software to edit those recordings on a computer can be found for free, software like Audacity, and there are lots of places on the web where you can upload this work, annotate it, and share it. *Sounding Out!* is a great example of this kind of work—they just use the WordPress platform, SoundCloud, and YouTube. It's not that those things are perfect by any means, but in terms of barriers to entry, they are very, very low. The main issue is that labor that would be compensated in publishing is volunteered in editing the site. They explain their practice as a "labor of pleasure" in their piece, but it does raise a bigger issue around the increasing concentration of tasks in the person of the scholar (which makes us a lot like artists and musicians, who are suddenly also simultaneously publishers, producers, promoters, etc.).

If you want to start doing big data analyses of an author's corpus, and that author's work hasn't already been digitized and you don't have access to a digital humanities lab, then that is a much more expensive proposition. For instance, the work that Tanya Clement is doing with HiPSTAS—that's a much more labor/capital/tech-intensive process that requires more advanced equipment (hardware and software), and technicians to work with it. And yet, as she talks about in her chapter, they are still having to figure out the basic, low-tech stuff, like how do you mark up audio in a way that is useful for scholars, and how do you actually analyze sound or get a computer to do it for you so that you can work at a higher meta level of interpretation? So there are a lot of dimensions to digital sound studies that are low-tech. If we want to follow my music research colleagues and start wiring up musicians to generate huge datasets based on their movements, that's going to be a lot more expensive, but those activities also don't mean much in a humanities context without rich humanistic questions to drive the inquiry.

WHITNEY TRETTIEN I would completely agree that a lot of the best work is low-tech. As we began putting together our website, we found ourselves pushing against the idea of using an all-encompassing content management system or developing a big new tool, and instead we kept coming back to

HTML, simple web technologies, and very basic, small-scale projects as a model for digital sound studies work. But then, what has changed? So often we pinpoint technology as the thing that's changing all of this—but here it seems like we're all agreeing that technology is not the primary engine of change in the academy. What is?

JS Well, I certainly don't think it's technologically driven at all. I think it's institutionally driven. One of the reasons that digital humanities has burgeoned is that there's money behind it. It's one of the only places in the United States that you can actually apply for and get a large grant to do humanistic work. In the U.S. there have been some interesting crossovers with library and information science and curatorial practice and preservation and things like that, so there are these huge institutional incentives to get into digital humanities.

There's also the logic of academic fashion. Digital humanities is a new thing—I mean, there are arguments about when it was coined and whatever, but the term isn't really in circulation before the twenty-first century—so it gets to be the new hotness. Every generation of scholars has to figure out how to get out of the intellectual mess made by the last generation. Through the eighties and nineties, it was the hermeneutic turn, the spatial turn, the theory moment. And now, instead of everything being about this hermeneutic turn, there seems to be this knee-jerk materialism that has replaced it. You can see it in the turn toward practice, of which digital humanities is a part. In Canada, there is a different-but-related practice called "research creation," or in parts of Europe it's called "artist research." It's tied to producing some kind of aesthetic work as the output of scholarship instead of a written piece. Often it comes out of an art school tradition, though, and some of it comes out of the need for artists to earn PhDs where the MFA used to be the terminal degree. But, as with digital humanities, it also represents a turn toward practice, and a very different response to the critiques of scholarly writing that came from our teachers and their teachers.

Any time you have this kind of ferment, it's an opportunity to ask real questions about how we do our work and what might be most useful. When you think about something like the journal article, that is a textual genre that changes about every quarter-century. The codex is much more durable, but the journal article is not a long-term thing that can't be messed with. So the digital humanities moment offers new opportunities to think about other kinds of periodical presentation of our work, especially when it comes to audio. It's child's play to put audio inside a PDF or inside a Kindle book or something like that; the only reason that it's not done is fear about copyright litigation. Our own unwillingness to fight for our fair-use rights, and bad old habits, are the only things that keep scholarship so silent.

It also remains hard to mark up audio. You can sort of do it with the Scalar video player, but it's inelegant. You can do it on SoundCloud, but of course SoundCloud isn't designed for scholars marking stuff up, and so it has these other dimensions and issues to it. Its social model isn't very good for scholars. It also has yet to turn a profit, which means the platform could change or disappear any day. There's Joanna [Annie] Swafford's Augmented Notes project, which is super cool, but it assumes that you're working with a musical score, and it's only really useful if written music enhances your argument. I'm struggling with this myself right now. I've got a piece on Auto-Tune that I'm almost finished with, and I'd like to publish it digitally. There are a few places that I want to annotate short audio clips and say, "Here's what we're talking about when we're talking about really audible pitch correction"; and "This is why this is Auto-Tune and not a vocoder in this track"; and stuff like that. I mean, I can do it in SoundCloud, I can do it in Scalar, but neither provides the kind of reading experience I want to offer my readers. So, on the one hand we do need better tools, on the other, we're pretty close in a lot of domains to being able to do a lot of stuff already. And most of the resistance as well as the impetus is institutional rather than having the tools.[1]

MARY CATON LINGOLD This might be a good moment to follow up on some of those institutional problems that you talked about. We initially tried to find an academic press to publish the web collection and found that presses were concerned about being able to manage the project within their ecosystems. They wanted us to use an existing platform, for example, but we argued that HTML would actually be much simpler and longer-lasting as a technology than most content management systems. In the end, we self-published the project, but now we're facing similar challenges preparing to archive the project so that it can be preserved at Duke Libraries. So it's been really interesting to see how libraries and publishers are thinking about the production of digital scholarship. In terms of archiving, websites are not pieces of paper that you can stick in a box, and there are legitimate institutional concerns about scalability. As a scholar invested in advocating for the value of multimedia scholarship, what do you have to say to the academic publishing world out there and to tenure and promotion committees about fixing this problem before this moment is gone? That's my fear: that there's money behind digital humanities now, but once this isn't the hot thing, are we going to lose some of that opportunity for innovation?

JS No, I don't think so, because there are other pressures. Right now, mathematicians, scientists, and some branches of academic medicine are in open rebellion against the for-profit science publishers, so it's not all on the humanities. It's part of a bigger movement.[2] There are a lot of things to note in your question: there's the whole publication and prestige part, the platforms, and preservation, which are all different things that all begin with P [*laughter*].

The platform problem is a real one. Just think about print publication and all the different formats that libraries have had to figure out. What do you do with the book that is too big to fit on the shelf? Well, it has to go somewhere else. What do you do with an unbound periodical? Well, it's got to go in a box. These are all things that librarians had to figure out how to catalog and manage. So in one sense we need to ask, "What kinds of digital containers can we legitimately be expected to maintain, and what range of things can exist inside of those?" A bunch of HTML pages that reference one another is probably pretty easy to keep going, but when you get into multimedia stuff or anything that's more heavily coded, it can start to be a problem.

With traditional publishing a lot of this stuff wasn't on the shoulders of the people doing the scholarship. The press had the people who did layout, binding, and shipping. But with your website, you're doing the binding, the layout, and the shipping (though probably someone else is handling the warehouse). That's a fundamentally different proposition. On the one hand it's another case of work that used to be done by others devolving into something that falls on the shoulders of academics who are asked to do it—I wouldn't say for free—but on top of their other jobs.

The preservation of multimedia materials is utterly puzzling. If you want to preserve video games, you've got to preserve the whole ecosystem of which they are a part. It's the same thing with any kind of multimodal scholarship that depends on a certain kind of platform or artwork. One way to think about it is that not everything has to last. Some interventions are of the moment. But so much scholarship doesn't work on that temporality. Timely interventions from a generation ago become influential arguments for reasons that the authors could not have foreseen. So I'm not real happy with the "let it all fade away" solution.

Relatedly, one of the really important questions is whether the author can abandon the project. Because if you look at the life cycles of intellectuals, there are many different kinds, but in almost every case, the way people advance in their intellectual development is to finish projects and leave them, rather than to have to come back and continuously maintain them. That is why you have librarians and archivists whose job it is to maintain things. So we need a system that allows that kind of intellectual abandonment.

My hunch as someone who studies standards and formats is that we're going to wind up with standardization and official formats. And that's why the publishers wanted to push you into using their platform. But of course their platforms change all the time! So it's not a very reassuring proposition at the moment. Every year I go back to my website and I update it, and there are links to the books that I've edited, coedited, and authored, and almost every year the web pages that I've linked to are no longer there and they've moved somewhere else. Lisa Gitelman says the 404 error is the most common page on the internet.

WT I want to ask as a related side note, do you see a viable role for self-publishing in the academy in the future?

JS In some ways, all academic publishing is self-publishing in the sense that you have a group of academics that get together as a group and decide to put something out—especially the journals that are curated and edited rather than going through the blind refereed thing. Lots of important humanities journals are edited by collectives. That's not that fundamentally different from a collective on a website deciding what to put up, except perhaps in terms of prestige politics.

I think there's certainly room for it and people do it. Blogging persists in various forms and remains useful to people. But there are limitations to self-publication. While I agree with many of Kathleen Fitzpatrick's critiques of peer review, I also think peer review serves a tremendously useful function.[3] Academics aren't always the best judges of our own work. One of the reasons why so much academic writing is hard to read is because we don't edit each other very well and we don't let ourselves be edited—people get so precious about their prose. For me, the thing that's exciting about something like *Sounding Out!* is that it's heavily edited and curated; it's not blind-refereed but it's certainly a kind of peer review that's prior to dissemination. On one level you could call it self-publishing because it isn't associated with a publisher, unless you call WordPress a publisher, which I guess they are in a certain sense. But, it's also not the same thing as me putting something out there on my blog.

The other problem is in how people are going to find things. Publishers serve an important curatorial role and a promotional one, too. I do my best book shopping every year at conferences, especially in the wake of the collapse of most good academic bookstores. So, I like the idea of nonprofit academic publishing. The thing to remember about university presses is that they exist to lose money, just only a certain amount of money every year. I think they serve a useful purpose. If they suddenly disappeared, we'd have a lot more work to do, and we'd have a much harder time producing and finding our best work. And of course there are all sorts of bad behaviors protected and justified by so-called blind peer review which shouldn't be allowed. Obviously there are many places we can improve, but I see self-publishing as part of a bigger ecology of publishing rather than a solution in itself.

MCL Well, I think you got to publishing and a little bit about preservation, what about the other p—prestige? Thanks for editing our questions, by the way in your response. Well done [*laughter*].

JS If you think about what makes publications matter, there's the idealistic version that we all want to believe, at least I hope we do, which is that publications matter when people read them. You want to be read; that's what matters. I feel that this is the real test of digital humanities work. If people produce things that are really useful to other people, they'll go find it. And they'll use it, and they'll cite it, and the fact that it wasn't in the *Journal of Highly Prestigious Things* isn't going to matter because the work will be influential on its own.[4] But of course, there are all sorts of cases where people evaluate your publications without ever having read them or heard them or seen them, depending on what they are, and that's where the whole prestige things comes in. So reviewers ask: "I haven't read this piece, but is that a good journal?" Like that would tell you anything—crap gets into even the "best" journals!

And that's where one of the big blockages is right now with multimodal scholarship. You see it in written tenure requirements—where they exist. You see it in the questions framed in tenure review letters; and you see it when hiring committees look at the CVs of prospective applicants. In a bad job market, I tell DH people to show that they can do a little of both: you show that you can play by the rules and then you do it the way that you want to do it as well, and that's probably the best that you can hope for. My job as someone reviewing a CV for a hire or tenure or whatever is to explain why and how digital work matters. For instance, people who write in TV studies will often publish in *Flow*, which is an online, multimodal periodical, so when people publish there and I'm reviewing their files, I'll explain that this is actually an important place for their work to come out and it will probably be read and taught more than this other journal article in a more traditional outlet. So part of it is a matter of people who are being called upon to make judgments making the right judgments and explaining stuff to committees. It's far from ideal.

The other way you go is for organizations to specify sets of "best practices," which is what MLA is trying to do right now.[5] That can work if you have enough motivated people who will then take those recommendations on board. In the humanities there's a lot that has to be overcome—the single-authored article or book is still seen as the most basic unit of scholarly production, and if your work suddenly turns collaborative, well, how's that going to be evaluated? Is a hiring or tenure committee going to understand that? Hopefully, people doing collaborative work get hired with the understanding that they are expected to do what they already do. But people also change what they do. And so, we need to work to build institutional structures and traditions that support more kinds of scholarship.

Part of it is just a matter of time and part of it is people citing each other, too. I think that's really important. It's interesting, for instance, to look at what digital projects are referenced in this book. Sharon Daniel and Erik Loyer's "Public Secrets" and Emily Thompson's "Roaring Twenties" are mentioned, although as of yet people are citing these pieces to say, "Hey, look, you can do cool things with sound studies in the digital domain."[6] The next step is for people to cite work because of what it says as opposed to "Hey, now you can . . ." Like any other scholarship, digital humanities work needs to be able to travel beyond its own scholarly community.

MCL I want to circle back a little bit to your point that what makes something matter is whether or not it gets read. What about whether or not something gets heard? I think this is a real problem for digital sound studies—and we talk a little about this in the introduction—people's reluctance to spend time listening. There are so many cool things to listen to on our web collection, for example, but when I show them off to people they say, "Oh, cool." But do they actually take the time to hear it? What do we need to do to get people to listen to digital scholarship? Is the burden on the creator to make it utterly compelling, or is it a larger cultural problem that needs to be addressed in a different way?

JS It's a huge challenge. Part of the problem is precisely the demand of duration. If you think about how people read scholarly books, there are those who start on page 1 and finish on page 400. They might do it because they are going to stand up in front of a class and talk about the text, or maybe they are really excited about the book, or it's really close to their area of expertise, or if you are reviewing the book, one hopes anyway, that they read the whole thing. But that's not the way it normally works. Normally, scholars don't read books cover to cover, from beginning to end. We can say that people shouldn't be engaging with scholarship superficially. But the reality is that we do it all the time when we are trying to write an essay and looking for a fact, or a way of talking about something, or a quote. The index at the back of a print book is a tacit acknowledgment that people don't read books from front to back. What would an index for academic soundworks look like? Think of it as a metadata problem. If the audio file were well tagged, you could find the part you need in the same way you can navigate a book. Then people could listen to the whole piece, or find parts as needed. As Jeremy Morris has shown, digital music didn't take off online until the metadata problem was solved—I'm not sure why we would expect anything different for digital scholarship.[7]

I'm also curious about music information retrieval as another way into audio files, but in the short run it probably will be of more use to answer specific questions, like, "Could you train a computer to hear music such that you could actually trace the diffusion of elements of style in popular music?" But we don't know if that's actually possible. One can imagine writing grants to study this sort of thing, studying it for years, and discovering that the answer is "no" [laughter]. But if you could do that, you could give a very different account of stylistic history, influence, and imitation, and other aspects of popular music history. That's classic humanities territory.

DM What you were just saying reminded me of something that I've been thinking about recently, which is this idea of close and distant reading, or big data versus microhistory, which we might kind of interpolate into close versus distant listening. It seems like with digital humanities there's always this tension between the big data and the minutiae, and with humanities scholarship, there is the tendency toward looking at small details and expanding outward. I think that has been a central question for us; this tension between the big and the small, the distant versus the close, has been something that's come up again and again and again.

JS Well, I mean one of the ironies when you're talking about sound is that both are more possible than they were before. To close-listen to something in 1990 at a university meant that you had to have a record player or a tape deck in a classroom, which was unlikely. If you wanted your students to listen to something, they had to go to a library to listen to it. You'd have to be pretty motivated if you weren't in the music department.

Close listening is a lot more possible today, and close analysis of audio is a lot more possible than it was even five or six years ago. That's equally exciting to whatever big data possibilities exist. The commercial world, of course, is much more into the corpus question, recommendation engines, and so forth. Look at the new Apple music interface and the way they're constantly trying to figure out how to refer bands and acts to one another.

DM Like the music genome project that was the basis for Pandora.

JS Yes. All of that is a kind of distant listening. Whether humanists want or need that kind of technology, or whether we can co-opt it for our purposes, I don't know. It depends on how flexible the technology is. Academics did really well at co-opting photocopiers and email, although email has co-opted us back now [*laughter*]. But sonically, we don't know what it would mean to use it to analyze a corpus. One of the challenges with that stuff is not to ask such conservative questions like, you know, getting a bunch of orchestral music and saying, "Why is this the best music that has ever been made?"

DM Yeah, tell me about it! Please, no more of that question.

JS Yeah, that's not a real research question, because you've basically said, "I want to use science to justify my aesthetic preferences." It's not going to happen. It rarely works in traditional humanities arguments—or at least I find those kinds of arguments completely beside the point of studying culture. It certainly isn't going to work when you need reproducible results. But the other challenge is that generating data itself is difficult. If you look at the brain science on hearing and music, a lot of it is done around very small sample sizes, because it's expensive to do brain scans. So, I don't know. I think we're actually still a pretty long way away from any real advances in this area, because even though the tech industry thinks in very short time horizons, stuff for us changes really slowly, at the intellectual level. At the blink of an eye an institution can change, obviously, but intellectually I think it takes longer. And so a lot of what we have to do is figure out what questions we can ask with digital tools that actually might be useful to answer. And I think when that happens, that's when these sorts of new methods will really take off.

WT A lot of what we've talked about amongst ourselves is that we're trying to bring sound into academic argumentation, and into academic practices of reading and writing—but in fact we need to bring academic practices of reading and writing a little bit more into sound in order to make audio mesh with scholarship.

JS Absolutely. Part of the problem is that there are not well-developed academic practices of listening outside of music and linguistics and a few other fields. Poetics is a really interesting example right now, because the field has become so much more sonically attuned in the last ten years—in part because of the online sound archives and in part because of all the digital humanities research around it, and the continued burgeoning of performance studies and its impact on literary studies. Poetry is really a place where, in the space of a generation, scholars have rediscovered the importance of listening and integrated it into their research and pedagogy. So it definitely can happen in other fields.

DM It seems like you're saying that the interplay between sound and text constantly finds ways of reinventing itself. People working on sound are always confronting the issue of writing about sound in text. But as we found, even when building a website dedicated to sound, we were constantly being forced to deal with the fact that a digital medium is a visual medium as well. On the one hand, sound studies is very good at critiquing this dichotomy between the linguistic and the aural; but I think at a different level, it's also not so much about a textual bias as it is about recognizing that design has these biases built in. What can we do from there, other than just point to them?

MCL For example, we felt frustrated by audio players being the primary mode for interacting with sound in digital spaces. It's kind of an analogue notion, that you have this box and it has a play button, and pressing it is how you hear sound—it disallows a more intuitive, deeply integrated way of experiencing audio. At the same time, if you just have sound bursting out of the speakers without any stimulus, it's really disruptive. Some of the more classic cultural biases a sound studies practitioner might address were very much embedded in the process of trying to design the website.

JS The tyranny of the player is a thing. I've been thinking about this too—about what the "intuitive" modes of sonic representation are.[8] There is the wave form, which is amplitude. There is the spectrum, which is pitch versus amplitude, or frequency versus amplitude, which is supposed to represent timbre, but no one seems to be able to figure that out. There's a very limited vocabulary for representing sound. If you go back fifteen years and you look at some of the really innovative work that was done in Flash by professional companies for band websites (most of it is no longer available)—they found all sorts of ways of representing music. Of course, they conformed to no standard other that the Flash standard, which meant the site had to load; and if we're talking about 2003 that took forever. But there was this moment of experimentation. And then people sort of gave up. There is a lot more power in HTML5 than in previous incarnations of the technology. You can basically build in plugins into your browser. So there is more that can be done. The kinds of vocal effects that are in Paperphone, the project by Umi Hsu and Jonathan Zorn in *Provoke!*—you can probably do that in a browser now.[9]

The player solved a problem, though. When sound became part of the internet, it immediately became annoying, because its first uses were for advertising, right? Annoying things just started to play when going to a site, which is a problem if you are in an office, or if you are in any kind of collective space. It violates the privacy that you imagine exists between you and your screen, even if we know it never does.

I don't have a ready-made, how-to answer for it, but it seems that there are many other ways of representing sound and we might try some. I think the player is useful and works well when the sound is an example in a piece, like a figure or an illustration. And as for the analog tape recorder reference, it's just classic skeuomorphic design combined with international standards. That right-facing triangle on the play button is part of an international standard and somebody somewhere did sit down and say "this means play" in all languages. Engineers and designers use it as a kind of semaphoric language. So I don't want to dismiss it either and say that an avant-garde strategy is automatically better, but it really depends on what you are trying to do.

WT Maybe we should turn to this collection more specifically. Are there any particular pieces that resonate with you, or did you notice any overarching themes or trends that you found interesting?

JS After reviewing the pieces, the first thing I did was make notes of all the different pedagogical suggestions that people have, because one of the great things about digital humanities work versus other fields is that people talk a lot more openly about pedagogy. I love teaching and I love talking about it, so it was actually really useful to see what others were doing in their classes. For the first time next winter [2016] I'm teaching a one-hundred-student undergrad lecture course in sound studies. I'm trying to figure out how many crazy things to do within the timeframe and labor structures. . . . What can I do that will actually work and not force the course to collapse under its own weight? What can the TA and I actually pull off? So the first thing I did when I finished reading the collection was open up my Evernote document that lists all the things I want to try in the class and I just added a bunch of suggestions from people's essays.

Zora Neale Hurston shows up a couple times (in essays by Myron Beasley and Regina Bradley). She's kind of hot right now. Daphne Brooks has written about her, Roshi Kheshti's new book also talks about her, and she keeps coming up at conferences I attend.[10] You always go back and reinvent your traditions, and she's now this really useful figure for a lot of different, newly invented traditions, whether we're talking about a sort of black feminist version of sound studies, or a digital humanities version of sound studies that's more based in practice—you don't just go out and record the songs, you also learn them yourself. And that's how you know tradition. Of course, ethnomusicologists have been doing this for a long time. What's different is that we're imagining it for sonic practices beyond music making or songs. So Hurston is interesting because she's a model of what's possible and also because her relation to her subjects was not the traditional ethnographic relationship of the time.

There's a real emphasis on experience. Steph Ceraso goes furthest in actually talking about body consciousness and the centrality of experience in listening. But there's a ton of that in the book implicitly, where people say, "I was only able to make sense of X because I experienced it in this way." So I think it's a really central-truth claim that's made a lot around multimodal scholarship, around its epistemic promise. But it's tremendously under-theorized. And Steph really went for it. Rich Rath does too. The great thing about Rich's piece is, and this is true of all his work, is this wonderfully tender attention to alterity. He works to think with the other but not by trying to be or inhabit the other. He's got that great line in the piece where he says there's no such thing as absolute slavery, that's a fantasy of the dominator, not the experience of the dominated. (I'm paraphrasing, of course.) That's a pretty powerful argument to draw from your work, and it's an interesting proposition. He sort of throws it out there because he's trying to explain what he was doing in terms of audio production and making music and how that ties into history scholarship. But I think ultimately it's arguments like that that we want to be pushing for in thinking about what digital humanities scholarship can offer a broad audience.

The only other obvious thing to point out is that most of the work discussed is collaborative—if not officially, then unofficially. You have people developing digital platforms, and even when they do it "by themselves," they do it with other people. It seems like there is a real emphasis on process and the value of actually doing stuff sonically. I think the challenge is to articulate that for people who don't already buy the argument, and I don't think anybody has succeeded yet. It's a hard thing to do. I'm not exactly sure what I would say to someone who asks, "Well, why should I bother with it?" I'm not sure I could convince you if you weren't already convinced, at least not without resorting to clichés that aren't actually true, like claims about sound and duration. But it's something we ought to think about. This is something I always push with sound studies in general. It isn't just, "Oh, hey, sound is great, let's study sound now"; but rather it's our job to contribute back to the big intellectual, philosophical, empirical, political questions that are challenging scholars across the humanities and social sciences. You guys are just trying to figure out what the hell this digital sound studies thing is, what digital humanities and sound studies might be together, what can we actually get done, what can we do. But the long game of it for me is how will this carry the big conversation forward, and what can we do. You know, how can we transform other people's minds.

MCL That's a good high watermark to aim for. That's great.

JS Yeah. I like ambitious [*laughter*].

MCL I really like that. So, I think that we take on a smaller task, which is to say, what sound studies brings to digital humanities and what digital humanities brings to sound studies, and why these two fields need to be in an explicit conversation with one another. I think the main thing that we think sound studies brings to digital humanities is an attention to culture, and I think what you were identifying in terms of Steph's theorization of how and why we learn differently through sonic experiences, and that being integral to all of this work—sound studies has a longer history of tying those kinds of insights to culture and history and really grounding them in, for instance, the history of technology, whereas digital humanities could do more to extend praxis into more deeply rooted humanistic research. And that's not a criticism so much as just something that sound studies nicely brings. Like you were saying, on the one hand, sound studies isn't just about, "Oh, sound's cool, we should study it." But on the other hand, sound *is* cool, and we *should* study it, and digital humanities could do more. There are implications for cultural productions that aren't text-bound, and you're reaching more diverse intellectual traditions when you open research up to the sounded world.

WT Similarly, along those lines, I think sound puts pressure on every single thing we've talked about, all the Ps: the prestige, publication, production, praxis, all of that. Sound brings something new to the conversation. And I think one of our goals was to try to demonstrate what that is. People are learning by playing with things that they've never played with before, things that they've never even been trained to address, and how do we bring that energy back into a traditional scholarly publishing economy. It's especially true for sound, because this book is silent, it's text-based—which gets back to this whole issue of how we bring listening practices into that, but also how do we bring reading and writing practices that are so well developed for good reasons within the humanities back to sound and sonic practice. I think that's where we see our intervention.

MCL So, Jonathan, are you jumping on the digital sound studies bandwagon?

JS Well sure, but with an asterisk. You say digital humanities is overwhelmingly visual, and I think absolutely, that's incontrovertible. Although of course there are great examples of sonic work in digital humanities, and people in the field know that it's an issue, too. So, obviously, there's that dimension of it. I think you're right that sound brings in different kinds of traditions, and different kinds of people—sound culture opens out into questions of race, gender, disability, and postcoloniality quite differently than visual culture. It can orient our research questions differently as well. But for me it is really driven by the questions rather than the methods or the tools. I can remember a time not so long ago when I said, "I will start using Powerpoint in my talks when I see five talks in a row with Powerpoint that doesn't fail." It was such a glitchy thing, and laptops were a lot less powerful. We've come a long way. But there's still more to be done. I have had to resort to a tech rider for my talks because, so often, basic audio setups don't work in the places I go, even when they are supposed to. Playing audio off a computer, while seated, while talking, is still a tough demand for many academic settings.

Perhaps unsurprisingly for someone who writes about technology, I am a bit of a gearhead. So it's only reasonable that my own practice has continued in a multimodal direction that started back in my *Bad Subjects* days. In my talks now I often use Ableton Live as a presentation software: I do audio editing, I do video, and I've used it to solve access problems, too. A few years back at the American Studies Association conference, I was on a sampling roundtable. It didn't seem like they were going to be able to get me the accommodation that I needed, because I have this vocal cord impairment, and it is difficult for me to stand up and deliver a talk. So I just recorded my talk beforehand and performed it using Live and my laptop. Since the talk was on sampling, I made it entirely of samples. I just stood up there and performed without speaking in the moment—I delegated my speech to the device. And it solved the problem. Everybody thought it was this kind of high-concept performance, but actually it was just an elaborate disability accommodation in an environment that otherwise could not accommodate vocal differences of that sort. When I finished it, I didn't know what to do with it. There was no obvious venue for something like that. But now, a revised version of that piece is going to come out as a publication in the online journal *Intermedialities*.[11] I'm happy to be able to "put out the single," as it were, even if it's a bit less fun than the live show.

But I will finish with my asterisk. When you say "digital sound studies bandwagon," which I realize is meant with some humor, it does raise a deeper concern for me. There are a lot of digital humanities bandwagons at the moment. But what we need are deep and multidimensional infrastructures. All the things we discussed in this interview are at their base infrastructural concerns. We need technical infrastructures to support the specific work we want to do, like tagging and marking up an audio file inside an electronic written text. We need institutional infrastructures to keep publications alive and running so their authors can abandon them. And we need cultural infrastructures where people develop and sustain more advanced techniques of listening to scholarship, as well as to the world, and where we better support one another's intellectual forays into sound. Given the choice, I'd rather get on the infrastructure than on the bandwagon.

NOTES

1 The Auto-Tune piece has been folded into my [JS] book with Mara Mills, *Tuning Time*. This book will include all sorts of historical audio—talking books, experimental time-stretching and time-compression recordings, pitch-shifting demos, snippets of musical works, modern examples, and we are still searching for a decent web audio player with markup for scholars. It does not seem to be anyone's priority.

2   For example, see the Cost of Knowledge (accessed November 30, 2017, http://thecostofknowledge.com). Resources for humanists to know their rights as authors include "Author Rights: Using the SPARC Author Addendum" (accessed November 30, 2017, www.sparc.arl.org/resources/authors/addendum). Publisher agreements are often littered with confusing—and sometimes illegal—legalistic language. As Sterne noted sometime after our conversation: "For example, recent contributor contracts from Princeton University Press and Palgrave have asked me to sign noncompete clauses (completely unnecessary), to warrant that no processes in my text could be harmful to readers trying to reproduce them (unnecessary), to give up my moral rights (not legally possible in Canada), and allow them to assign the work to another author for revision and republication (just plain asinine)." Also see Striphas, "Acknowledged Goods."
3   Fitzpatrick, *Planned Obsolescence*.
4   This is not an actual journal—at least not yet.
5   See MLA, "Statement on Electronic Publication" (accessed November 30, 2017, www.mla.org/statement_on_publica).
6   Thompson, "The Roaring Twenties," and Daniel, "Public Secrets."
7   Morris, *Selling Digital Music*.
8   See Sterne, "Player Hater."
9   For more on Paperphone, see http://soundboxproject.com/project-paperphone.html (accessed January 15, 2018).
10 Brooks, "Sister, Can You Line It Out?," and Khesti, *Modernity's Ear*.
11 Sterne, "Through the Fog of Sonic Memory."

WORKS CITED AND FURTHER READING

Brooks, Daphne. "Sister, Can You Line It Out? Zora Neale Hurston and the Sound of Angular Black Womanhood." *Amerikastudien/American Studies* 55, no. 4 (2010): 617–27.

Daniel, Sharon. "Public Secrets." *Vectors: Journal of Culture and Technology in a Dynamic Vernacular* 2, no. 2 (2007): n.p. http://vectors.usc.edu/projects/index.php?project=57.

Fitzpatrick, Kathleen. *Planned Obsolescence: Publishing, Technology, and the Future of the Academy*. New York: New York University Press, 2011.

Khesti, Roshanak. *Modernity's Ear: Listening to Race and Gender in World Music*. New York: New York University Press, 2015.

Morris, Jeremy. *Selling Digital Music: Formatting Culture*. Berkeley: University of California Press, 2015.

Sterne, Jonathan. "Player Hater." *Flow* 15, no. 2 (October 30, 2011): n.p. http://flowtv.org/2011/10/player-hater.

Sterne, Jonathan. "Through the Fog of Sonic Memory: Granulation, Groove Extraction, MIDI Conversion, FSU." *Intermédialités* 23 (Spring 2014): n.p. www.erudit.org/revue/im/2014/v/n23/1033341ar.html?vue=resume &mode=restriction / doi:10.7202/1033341ar.

Striphas, Ted. "Acknowledged Goods: Cultural Studies and the Politics of Academic Journal Publishing." *Journal of Communication and Critical/Cultural Studies* 7, no. 1 (2010): 3–25.

Thompson, Emily. "The Roaring Twenties: An Interactive Exploration of the Historical Soundscape of New York City." *Vectors: Journal of Culture and Technology in a Dynamic Vernacular* 4, no. 1 (Fall 2013): n.p. http://vectors.usc.edu/projects/index.php?project=98.

CONTRIBUTORS

MYRON M. BEASLEY, PhD, is associate professor at Bates College, where he teaches in the areas of American studies, African American studies, and gender and sexuality studies. His ethnographic research includes exploring the intersection of cultural politics, material culture, and social change. He has been awarded fellowships and grants by the Andy Warhol Foundation, the Whiting Foundation, the Mellon Foundation, and the National Endowment for the Humanities, and, most recently, the Ruth Landes Award from the Reed Foundation. His ethnographic writings about Africana cultural politics, contemporary art, material culture, and cultural engagement have appeared in many academic journals, including *Text and Performance Quarterly*, *Gastronomica*, *Journal of Poverty*, *Museum and Social Issues*, *Journal of Curatorial Studies*, *Food and Foodways*, and *Performance Research*. His food–film/installation–ritual/feast (of his ethnography in Brazil) has appeared in the UMMI and Paris film festivals. His recent curatorial projects include the Ghetto Biennale (Haiti), CAAR Paris 7 (France), and Dak'art (Senegal).

REGINA N. BRADLEY is a writer, scholar, and researcher of African American life and culture. She is an alumna Nasir Jones Hiphop Fellow (Harvard University, spring 2016) and an assistant professor of English and African diaspora studies at Kennesaw State University in Kennesaw, Georgia. Her expertise and research interests include hip-hop culture, race and the contemporary U.S. South, and sound studies. Dr. Bradley's current book-length project, *Chronicling Stankonia: OutKast and the Rise of the Hip-Hop South* (under contract, University of North Carolina Press), explores how Atlanta hip-hop duo OutKast influences conversations about the black American South after the civil rights movement. She can be reached at www.redclayscholar.com.

STEPH CERASO is an assistant professor of digital writing and rhetoric in the English department at the University of Virginia. Her research and teaching interests include multimodal composition, sound studies, pedagogy, digital rhetoric, and sensory rhetorics. Ceraso's article, "(Re)Educating the Senses: Multimodal Listening, Bodily Learning, and the Composition of Sonic Experiences," won the 2015 Richard Ohmann Award for Outstanding Article in *College English*. In addition to coediting a "Sonic Rhetorics" issue of *Harlot*, she has published scholarship in *Composition Studies*, *Currents in Electronic Literacy*, *Sounding Out!*, and *Provoke! Digital Sound Studies*. Ceraso's book, *Sounding Composition* (University of Pittsburgh Press, 2018), proposes an expansive approach to teaching with sound in the composition classroom.

TANYA E. CLEMENT is an associate professor in the School of Information at the University of Texas at Austin. She has a PhD in English literature and language and an MFA in fiction. Her primary area of research centers on scholarly information infrastructure as it impacts academic research, research libraries, and the creation of research tools and resources in the digital humanities. Clement's digital projects include High Performance Sound Technologies for Access and Scholarship (HiPSTAS), which focuses on developing a virtual research environment in which scholars and cultural heritage professionals can better access and analyze audio collections with machine learning and visualization.

REBECCA DOWD GEOFFROY-SCHWINDEN is an assistant professor of music history in the College of Music at the University of North Texas. Her research has been presented at annual meetings of the American Musicological Society, Society for Ethnomusicology at the Congress of the International Musicological Society, and at conferences in the United States, France, Portugal, and Finland. Her publications have appeared in *Kinetophone: Journal of Music, Sound, and the Moving Image*; *Studies in Eighteenth-Century Culture*; and *Women and Music: A Journal of Gender and Culture*. Rebecca holds BAs in history and international studies, phi beta kappa, from Pennsylvania State University's Schreyer Honors College. She earned an MA and PhD in musicology from Duke University, where she was inducted into the Society of Duke Fellows.

W. F. UMI HSU is an ethnomusicologist and sound ethnographer working on public humanities and civic design projects. Currently a digital strategist and impact researcher at the City of Los Angeles Department of Cultural Affairs, they are the founder of Lab at DCA, where they provide research and strategy to redesign the department's data and knowledge architecture. As a researcher, Hsu has done fieldwork to uncover the politics of street music-culture in Taipei. They also lead LA Listens, a sound-based community engagement project, and cofounded Movable Parts, a maker collective that reimagines public spaces in Los Angeles. Umi Hsu has a PhD in critical and comparative studies in music from the University of Virginia, and they have taught at Occidental College, Art Center College of Design, the University of Southern California, and the University of Virginia.

MICHAEL J. KRAMER works at the intersection of history, the arts, cultural criticism, public humanities, and digital technologies. He is the author of *The Republic of Rock: Music and Citizenship in the Sixties Counterculture* (Oxford University Press, 2013; paperback, 2017). Kramer's current research explores the relationship between technology and tradition in the U.S. folk music revival; it includes a multimodal project about the Berkeley Folk Music Festival, which took place annually at the University of California between 1958 and 1970, as well as more technical research in digital humanities methods. Kramer is associate director of the Digital Liberal Arts at Middlebury College, where he teaches history and American studies. He previously taught at Northwestern University, where he cofounded the institution's Digital Humanities Laboratory. He serves as dramaturg and historian-in-residence for The Seldoms dance company and writes for numerous publications as well as for his own website, michaeljkramer.net.

MARY CATON LINGOLD is an assistant professor of English at Virginia Commonwealth University. Her research explores the literature and culture of American slavery and colonization, early Afro-Atlantic music, sound studies, and digital humanities. Her digital projects include Musical Passage: Voyage to 1688 Jamaica (www.musicalpassage.org) and the Sonic Dictionary (www.sonicdictionary.org). She has published in *Early American Literature* and *American Literature*, and she is currently writing a literary history of Afro-Atlantic music (1650–1850).

DARREN MUELLER is an assistant professor at the Eastman School of Music at the University of Rochester. His research interrogates how technologies of sound alter the development of musical styles and the means by which music making enables cultural agency. His current book project, tentatively titled *At the Vanguard of Vinyl: A Cultural History of the Long-Playing Record in Jazz*, details how musicians and other industry professionals leveraged the rapid rise of mass consumption in the postwar era to improve jazz's cultural positioning within the Unites States. His work has appeared in *Jazz Perspectives* and the *Journal of the Society for American Music*.

RICHARD CULLEN RATH is associate professor of history at the University of Hawaiʻi at Mānoa and director of the Digital Arts and Humanities Initiative of the College of Arts and Humanities. He is the author of *How Early America Sounded* (Cornell University Press, 2003) and is currently working on two books, one an introduction to the history of hearing and the other comparing the rise of print culture in the eighteenth-century Atlantic world to the rise of internet culture today. He has also written three award-winning articles on music, creolization, and African American culture. Recently, he has published articles on silence and noise, the sonic dimensions of wampum, media and the senses in the Enlightenment, and the open-source digital future of the humanities. He teaches courses on digital humanities, the history of media and the senses, sound studies/history of hearing, and Atlantic history. Rath is also a musician who has found ways to use music to "do" history, sound, and American studies whenever possible.

LIANA SILVA is a high school teacher, editor, and writer living in Houston, Texas. She obtained her PhD from the English department at SUNY Binghamton. Her dissertation, *Acts of Home-making*, is a study of how African Americans and Puerto Ricans represent New York City as a home. She has written for the *Inside Higher Ed* blog, University of Venus, Chronicle Vitae, and the *Houston Chronicle*'s Gray Matters. In the past she worked as a graduate writing specialist at the University of Kansas Writing Center and as a graduate writing instructor at the University of Texas Health Sciences Center. Liana is currently working on a book about postcards, stemming from her geeky obsession with quirky postcards. You can find her on Twitter at @literarychica.

JONATHAN STERNE is James McGill Professor of Culture and Technology in the Department of Art History and Communication Studies at McGill University. He is author of *MP3: The Meaning of a Format* (Duke University Press, 2012); *The Audible Past: Cultural Origins of Sound Reproduction* (Duke University Press, 2003); and numerous articles on media, technologies, and the politics of culture. He is also editor of *The Sound Studies Reader* (Routledge, 2012) and a coeditor of *The Participatory Condition in the Digital Age* (University of Minnesota Press, 2016). His current projects consider instruments and instrumentalities; mail by cruise missile; and the intersections of disability, technology, and perception. His next book, tentatively titled *Tuning Time: Histories of Sound and Speed*, is coauthored with Mara Mills. Visit his website at http://sterneworks.org.

JENNIFER LYNN STOEVER is author of *The Sonic Color Line and the Cultural Politics of Listening* (NYU Press, 2016). She is also cofounder and editor in chief of *Sounding Out!* and associate professor of English at SUNY Binghamton. She has published in *American Quarterly, Social Text, Radical History Review*, and *Modernist Cultures*, among others. Her latest research on black and Latinx women record collectors and their role in early hip-hop is forthcoming in the *Oxford Handbook of Hip Hop Studies*.

JONATHAN STONE studies rhetoric within and across its historical, cultural, and vernacular contexts. His current research is focused on the sonic rhetorics of American vernacular music in the 1930s. His book project, *Listening to the Lomax Archive*, investigates the careers of John A. Lomax and his son Alan during the Great Depression, with focus on field recordings made for and stored by the Library of Congress's Folklife Archive. Generally, Jon is interested in the mythologies that surround the notion of technological advance, particularly as such narratives reveal the tensions and rhetorics at the intersection of "traditional" and "progressive" ways of thinking and being. He is currently an assistant professor at the University of Utah.

JOANNA SWAFFORD is the Digital Arts and Humanities Specialist at Tufts University. Her articles have appeared or are forthcoming in *Debates in Digital Humanities, Journal of Interactive Technology and Pedagogy, Victorian Poetry*, and *Victorian Review*. She is the project director for Songs of the Victorians (www.songsofthevictorians.com), Augmented Notes (www.augmentednotes.com), and Sounding Poetry; she is also head of pedagogical initiatives for NINES (Networked Infrastructure for Nineteenth-Century Electronic Scholarship, www.nines.org). Before joining Tufts, she taught English and digital humanities at SUNY New Paltz.

AARON TRAMMELL is an assistant professor of informatics at the University of California, Irvine. He earned his doctorate from the Rutgers University School of Communication and Information in 2015 and spent a year at the Annenberg School of Communication at USC as a postdoctoral researcher. Aaron's research is focused on revealing historical connections between games, play, and the U.S. military-industrial complex. He is interested in how political and social ideologies are integrated in the practice of game design and how these perspectives are negotiated within the imaginations of players. He is the editor in chief of the journal *Analog Game Studies* and the multimedia editor of *Sounding Out!*

WHITNEY TRETTIEN is assistant professor of English at the University of Pennsylvania. Her research spans the fields of Renaissance literature, book history, women's studies, digital humanities, media archaeology, and sound studies. Whitney is currently working on a hybrid print/digital monograph titled *Cut/Copy/Paste: Fragments of History*, which explores the relations between fragments, history, books, and media in seventeenth-century England. Whitney is also the codesigner and coeditor of *thresholds*, a digital journal for creative/critical scholarship. Before joining the faculty at Penn, she taught at the University of North Carolina, Chapel Hill.

INDEX

academic prestige, 273–74
academy (the), 3, 10–11, 83. See also Sterne, Jonathan, interview
acoustemologies, 35, 232–33. See also audible history; historical acoustemologies
acoustic design, 257–57
acoustics (audible history), 235–38, 240
Adaptive Recognition with Layered Optimization (ARLO), 157–58, 160–61, 165
Advanced Research Consortium, 170
affect studies, 37
African American cultural production, 40, 48–49. See also black studies; blues music; Hurston, Zora Neale; OutKast; "OutKasted Conversations"
"America Eats" project, 57–58
archaeoacoustics, 235, 245n3
archives: "America Eats," 57–58; of animal sounds, 18n22; audible history projects creating, 244; Berkeley Folk Music Festival, 14, 180; of bodily sounds, 53; of digital humanity, 66, 69, 72; digitizing, 8; English Broadside Ballad, 217; folk life, 73; Hurston's work in, 12, 57–58; Jazz Loft Project, 1–4, 11, 16n1; and memory, 72; and orality, 72–73; "OutKasted Conversations" as, 13, 69, 122, 128; PennSound Poetry, 157–59; performance as, 18n24, 52; personal sound, 232; pre-recording era, 233–34, 238, 245n4; "Public Secrets" project, 11, 257–59, 264n18; Radio Haiti, 8; repatriation of, 8; social sound, 232; Songs of the Victorians, 218–19
audible history: acoustics, 235–38, 240; archives, projects as, 244; auditioning subjects, 235–42; augmented reality, use in, 240–41, 245n3; bells example, 242; diverse media formats in, 232–33; earwitness accounts, 234–36, 239, 241–42; historically informed performance (HIP), 233; methods of, 232, 234; "Organs of the Soul" project and, 239–40, 242–43; pre-recording-technology era, 233–34, 238, 245n4; Projet Bretez, 240–43; and reenactment, 233, 237–38, 240, 242–44, 245n5; "The Roaring Twenties" project and, 10–11, 237–38, 240; roots of, 233; and soundscapes, 235–38, 240; strengths of, 244; technology, importance to, 233–34; transdisciplinary nature of, 243–44; Virtual Paul's Cross Project and, 235–37, 240, 246n13. See also historical acoustemologies
The Audible Past, 6, 203, 208n57
audio mark-up, 268, 270, 275
audio player interface metaphor, 277–78
auditioning subjects (audible history), 235–42

Augmented Notes: about, 219, 270; advantages of, 224; development of, 223–24; features, 219–20, 222; future directions, 222–23; goal of, 218; origins and inspirations, 215–17; Songs of the Victorians project, 218–19; strengths of, 222

augmented reality, 240–41, 245n3

Auto-Tune, 270, 282n1

Baldwin, James, 51

bamboomba (instrument), 35–36, 41

Berkeley Folk Music Festival: archive, 14, 180; duration of, 179; Hurt, Watson, and Hinton at, 182–85, 187; ideology of, 184–85; Lipscomb at, 179–81, 199; regionalism in, 187

Binghamton University Sound Studies Collective, 88, 90

blackface minstrelsy, 18n23

black studies, 5, 7, 18n23, 123, 129. *See also* African-American cultural production; Hurston, Zora Neale; "OutKasted Conversations"

blogging, 83–84, 91. *See also Sounding Out!*

blues music: OutKast, influence on, 122; as performative text, 53–54; song styles, debates on, 207n31; "unholy trinity," 53–54; Watson, influence on, 190; West African influence on, 34–35. *See also* Hurston, Zora Neale; Hurt, "Mississippi" John

Capturing Sound: How Technology Has Changed Music, 216

chapter overviews, 12–16

Choir! Choir! Choir!, 64–66, 71, 76

classification systems: about, 156; accuracy *versus* computational tractability, 164–65; encoded relationality, strengths of, 170–71; FADGI standard, 167, 170; fixed *versus* emergent meanings, 168–69; fixed *versus* fluid representations, 169–71; fixity in, 169; frames of reference, use of, 169; future work, 171–72; in HiPSTAS, 158–60, 163–64; IASA standard, 167, 170; and paralinguistic features, 161–63, 165–67, 171; and prosody, 161–63, 165–66, 168, 171; as sociotechnical phenomena, 156, 161, 164; as subjective, 156, 165–66, 168; text-sound compromise, 165–67. *See also* TEI's Transcriptions of Speech classifications

close listening, 4, 14, 139, 168, 275–76

Collins, Phil, 64–65, 67–68, 74–76

colonialism: and documenting others, 18n23, 144; and power, 7, 30; and source bias, 38; *Voyage to the Islands* and, 29–30, 32–35, 38

computational tractability, 164–65

copyright, 2–3, 270

craft production, 91, 98

creolization, 41

critical making, 139

cultural frameworks, 125–26

cultural memory, 6, 66, 73–74, 126

cultural studies, 5–6, 16n2, 49, 52, 124, 127

culturonomics, 9

data fusion (image sonification), 182, 193–95

data sonification (image sonification), 196–202

deformative making, 137

design, universal, 15, 253–55

digital activism, 91, 98

digital audio: accessibility of, 255, 261; annotation, need for, 156; Eurocentrism in, 38–39; history of, 31; spatial affordances of, 257; tactile experience of, 259, 261

digital humanities: in academia, 267, 269, 271; access, issues of, 262n3; blogging, view of, 84–85; and digital aesthetics, 202; Digital Humanities Summer Institute, 223; diversity, lack of, 93, 256,

263n14; infrastructures, need for, 282; and libraries, 9; low-tech solutions, 268; maker culture in, 84; materiality in, 134; overview of, 8–9; pedagogy, 278; peer review in, 14; reflexivity concept in, 137, 149n19; scholarship, 269, 273–74; sound, bias against, 9–11; sound, treated as semiotic, 251; and *Sounding Out!*, 84–85, 93, 101, 103; and sound studies, 8, 11, 16, 48, 267, 280; Sterne interview, 267–70, 282; text-centricity of, 9–11; visual, dominance of the, 10, 181, 281. *See also* digital image sonification; High Performance Sound Technologies for Access and Scholarship; sound practices

digital humanity: definition of, 66, 68; and digital vernacular, 67; and epideictic, 69, 75; evidence for, 67; folkness of, 69, 75; and libraries, 72; platforms enabling, 71; as secondary orality, 70–72; smartphone apps, 71; SoundCloud, 71; and vernacular digital culture, 67

digital image sonification: data fusion, 193–95; data sonification, 196–202; and digital aesthetics, 202; and evidence, 207n45; and historical consciousness, 182–83, 196–97, 202–4; materiality, as emphasizing, 203; scaling, 201–2; technique overviews, 182; tools for, 207n42. *See also* digital sound design

digital sound design (image sonification): about, 183–84; and historical consciousness, 184, 189–90; Hurt, Watson, and Hinton photograph, 182–83; Hurt, Watson, and Hinton project, 184–90; technology, advantages of, 184

digital vernacular, 65–67, 71

digitization. *See* remediation

disability, 253–55, 260, 262n3, 281–82

diversity, 93, 256, 263n14

DJs, 7, 48–50, 58

drums, in West African music, 33

earwitness accounts, 234–36, 239, 241–42

emoti-chair, 260–61, 264n24

English Broadside Ballad Archive (EBBA), 217

epideictic rhetoric, 69, 75

ethnodigital sonics, 30–32, 42. *See also Voyage to the Islands*, re-creation of music from

ethnography: and knowledge transmission, 137; and learning, 138–39, 141, 144; performance, 55; recording, politics of, 144–45. *See also* Hurston, Zora Neale

ethnomusicology, 18n24, 30–31, 38–39, 205n13, 233

"ethno-" prefix, 30

feather-bed resistance, 54

Feld, Steven, 7–8, 18n25, 232

field recordings: and learning, 142; politics of, 139, 144–45; repatriation of, 8; in "A Tale of Two Soundscapes," 254. *See also* Hurston, Zora Neale

folklore preservation, 54–56, 60n27. *See also* Hurston, Zora Neale

folkness: definition of, 69; of digital humanity, 75–76; rhetorical, 75–76; as secondary orality, 73–74; and sound, 69, 74; technology enabling, 71

folk revival, American: about, 73; Humbead's *Revised Map of the World*, 191–95; melograph, Seeger's, 205n13; recording aesthetic, 206n23; as secondary orality, 73–74. *See also* Berkeley Folk Music Festival; digital image sonification

Frazier, E. Franklin, 40–41

Freedom's Ring project, 11

hauntology, 50, 58

hearing: data, 182; as having history, 31–32, 235; instruments of, 171; as social practice, 238

Herskovits, Melville, 40–41

INDEX · 293

High Performance Sound Technologies for Access and Scholarship (HiPSTAS): about, 156–57, 172nn4–5, 268; and ARLO, 157–58; classification problems, 160; classification schema (TEI's), 158–60, 163–64; goals of, 157; machine learning, use of, 157–58; tagging interface, 158–59; and vocal gestures, 159; voice qualities, use of, 164

Hinton, Sam, 182–88

historical acoustemologies: digital approaches to, 239–44; recreation of, 238; reenactment of, 232–33, 238; and "The Roaring Twenties" project, 11–12, 237–38, 240; soundscapes and subjectivities, 235. See also audible history

historical consciousness, 182–83, 196–97, 202–4

historical imagination, 32–33

historical reenactment, 233, 237–38, 240, 242–44, 245n5

history, 31–33, 232. See also audible history; digital image sonification; historical acoustemologies

humanities: collaboration in, 274; current transition moment, 3–4, 9, 48; editing and peer review, 104; multimodal scholarship in, 15; and "re-" prefix, 48; sound studies in, 168, 280

Humbead's Revised Map of the World, 191–95

Hurston, Zora Neale: audio work as scholarship, 56; Beasley, influence on, 51, 56, 58; ethnography, influence on, 52, 124; feather-bed resistance, 54; field recordings, 55–57; fieldwork experiences, 54–55; fieldwork performance for Boas, 51–53; influence of on "OutKasted Conversations," 123–24; performance, strategic use of, 51–53, 58; as performance ethnographer, 50–52, 55–56, 123–24; personal life, mysteries around, 57–58; as postmodern, 52–53; recordings by, digitized, 57–58; in Red Rooster remix, 48–50; 58; in Sterne interview, 279; as "unholy trinity" member, 53; unmarked grave of, 57; vocal performances of, 53–56, 58; work attributed to others, 12, 57–58; writing of, 51, 56–57

Hurt, "Mississippi" John, 182–89, 207n32

I am I be concept, 49–50, 54, 56
immersion, 256–59
inclusivity, 252–56

Jazz Loft Project, 1–4, 11, 16n1
Journal of the American Musicological Society, 216

learning: assessment of, 131–32, 147; banking model, 132; codification of, 131; creative projects, 146; encoding and decoding culture, 142; ethnography, 138–39, 141; as interpretive act, 130; metrication of, 132; and reflexivity, 137–39, 149n19; and remediation, 134–37; resonance, 139–45, 147; and sense hierarchies, 131, 146; silencing of, 132–33, 147n2, 148n4; and sonification, 135–36, 146; sounds' potential for, 133–34, 145. See also pedagogy

Left of Black, 123

libraries: and digital humanities, 9, 269; and digital humanity, 72; and digital scholarship, 270–71; "folk-life" archives in, 73; recording studios in, 72; as silent spaces, 132

Lindsey, Treva, 126

Lipscomb, Mance: about, 179; photograph of at Berkeley festival, 178–81; sonification of Lipscomb Berkeley photo, 197–201

listening: academic, 277; close, 4, 14, 139, 168, 275–76; components of, 238; creating cultural frameworks, 125–26; to data, 196–97, 202, 208n51; distant, 276; embodied, 6, 250, 252–53, 279; empathetic, 144; and historical imagination, 32; historically situated, 235, 240; and learning, 131, 134–35, 143, 145; reluctance toward, 274; shared, 141–43; technology shaping, 6; and vibration, 260

machine learning, 158
materiality, 4, 7, 134, 203, 244
media ecology, 69–70, 77n12
media studies, 6–7
melograph, 205n13
methods: of audible history, 232, 234; of historical imagination, 32–33; Sounding Out! survey, 101–2
metrication of learning, 132
microtones, 35, 37–39, 41
multimedia preservation, 271
multimodal scholarship: and academic career paths, 273; accessibility of, 254; citing, 274; in humanities, 15; institutional support for, 15; maintenance of, 271–72; "Organs of the Soul" project as example of, 239–40, 242–43; preservation of, 271; Projet Bretez as example of, 240–43; "Public Secrets" project as example of, 11, 257–59, 264n18; "The Roaring Twenties" project as example of, 11–12, 237–38, 240; Romantic Era Songs Project as example of, 225n8; and Scalar platform, 7, 11, 14, 19n38, 217, 239, 270; Songs of the Victorians project as example of, 218–19; in sound studies, 12; Virtual Paul's Cross Project as example of, 235–37, 240, 246n13. See also digital image sonification; "OutKasted Conversations"; Sounding Out!; Voyage to the Islands, re-creation of music from
music: Angolan, 30, 32–34, 36, 39–40; in black studies, 5; Capturing Sound: How Technology Has Changed Music, 216; Choir! Choir! Choir!, 64–66, 71, 76; and digital technologies, 7, 31; ethnomusicology, 18n24, 30–31, 38–39, 205n13, 233; Eurocentrism in, 38–39; information retrieval, 275; Koromanti, 30, 33–37, 40; Norton anthologies of, 216; OutKast, 120–23, 127; popular, 5; and recording technologies, 7; Voyage to the Islands, 29–30, 32–35, 38; West African, 33–34. See also Augmented Notes; Berkeley Folk Music Festival; blues music; folkness; folk revival, American; Hurston, Zora Neale; "OutKasted Conversations"; Voyage to the Islands, re-creation of music from
Musical Instrument Digital Interface (MIDI): about, 31, 43n15; microtones in, 39; in re-creating music from Voyage to the Islands, 36–37, 41–42
Musical Passage: A Voyage to 1688 Jamaica, 43n2
Music Encoding Initiative (MEI), 222, 225n16

noise: class lesson on, 135–37, 148n12; in image sonification, 198–99; subjectivity of, 5, 237
Norton music anthologies, 216

Omeka, 9
Ong, Walter J.: contemporary relevance, 75–76; criticisms of, 68, 72–73; on folk life, 73–74; on literacy, 67–68, 72–73; and media ecology, 70; on oral-formulaic composition, 72–73; on orality, 67–69; as Phil Collins, 67; relevance of, 75; secondary orality theory, 70–72; in sound studies, 66–67; on technology, 76
oral interpretation, 51–52
orality, secondary: cycles of, 74; digital humanity as, 70–72; folk revival as, 73–74; Ong's theory of, 70–72
orality theory, 37, 72–73
"Organs of the Soul" project, 239–40, 242–43
OutKast, 120–23, 127

"OutKasted Conversations": about, 120, 123, 125, 128–29; advertising of, 128; as archive, 13, 69, 122, 128; digital format, advantages of, 122–23, 126–28; editing, 127; Hurston's influence on, 123–24; interviewee selection, 125; "Left of Black" influence on, 123; Lindsey's interview, 126; and listening, 125–26; origins of, 121–22; premise for, 122; and southern blackness, 123–25, 129; success of, 128; and viewer feedback, 127–28

Partridge, Dan, 1–3
pedagogy: in digital humanities, 278; Digital Music-Cultures course, 133–34, 148n5; as interpretive acts, 130; remediation's value in, 135; software design, similarities to, 130; in Sounds of Learning project, 138–145; Sterne on, 278–79. *See also* learning
peer review: advantages to, 273; in digital humanities, 14; disadvantages to, 104–5, 273; *versus* editorial processes, 105, 107; public, 127; Sterne on, 272–73
PennSound Poetry archive, 157–59
performance: as analysis, 58; as archive, 18n24, 52; ephemerality of, 49–50; ethnographic, 55; in ethnomusicology, 18n24; generative nature of, 50; and hauntology, 50, 58; historically informed, 233; as inquiry, 48; traditions of, 7, 18n23–24; of written texts, 51–52. *See also* DJs
personal sound archives, 232
photographs: of Hurt, Watson, and Hinton, 182–85, 187–88; of Lipscomb, 178–79; as media, 181; remediation of, 180–81. *See also* digital image sonification
Photosounder (software), 198–201, 207n42
physical experiences, 250–52, 259–61
platforms: academy, as the new, 114; changeability of, 272; enabling digital humanity, 71; media studies research on, 7; Omeka, 9; problems with, 270–72; publishing, 216–17; Scalar, 7, 11, 14, 19n38, 217, 239, 270; silence of, 10; *Sounding Out!* as, 95–101, 105. *See also* Augmented Notes
Pleasure Ninjas Collective, 126
poetry, 157–60, 164, 277
polyrhythm, 37–39
power: colonial, 7, 30; in cotton industry, 188; and digital humanities, 9; of historical imaginings, 33; of photographs, 199; sounds as, 140; and technology, 8. *See also Sounding Out!*: politics of
pre-recording era, 233–34, 238, 245n4
Projet Bretez, 240–43
prosody, 161–63, 165–66, 168, 171
Provoke! Digital Sound Studies, vii–viii, 66, 253, 268
"Public Secrets" project, 11, 257–59, 264n18
publishing platforms, 216–17

Radio Haiti archive, 8
recording technologies: in audible history, 234; constraints of, 18n24; for home computers, 31, 36; Hurston's use of, 55, 58, 124; in libraries, 72; in ornithology, 18n22; political nature of, 7–8, 18n23, 144–45, 181, 234
Red Rooster, 47–48
reflexivity (learning concept), 137–39, 149n19
remediation: definition of, 134; of digital images, 180, 197; and distortions, 181, 204n7; and learning, 134–37; pedagogical value of, 135; of photographs, 180–81. *See also* digital image sonification
"re-" prefix, 48, 59
research creation, 269
resonance (learning concept), 139–45, 147
rhetoric, 65–67, 73–75, 77n4
"The Roaring Twenties" project, 11–12, 237–38, 240
Romantic-Era Songs project, 225n8

sansas, 34–35, 37, 41, 43n15
Scalar platform, 7, 11, 14, 19n38, 217, 239, 270
secondary orality. *See* orality, secondary
Seeger, Pete, 65, 205n13
self-publishing, 272–73
semiotics, 251
sense hierarchies, 131, 146
slavery, 40–41
Sloane, Hans, 29, 34, 43n12. *See also Voyage to the Islands*
Smith, Eugene W., 1–2, 7
software design, 130
Songs of the Victorians project, 218–19
sonification, 135, 208n51. *See also* digital image sonification
SoundCite, 217
SoundCloud, 10, 71, 268, 270
Sounding Out!: blog format, advantages of, 91–92, 95, 97–98, 116nn16–17; blog format, consequences of, 91, 98, 114; community building approach, 86, 90–91, 105, 108; as community hub, 103–4, 111–12; as craft, 91, 98; as digital activism, 91, 98; and digital humanities, 84–85, 93, 101, 103; editorial process of, 104–8; founding of, 86, 88, 90, 100; as interdisciplinary, 91–93; as labor of pleasure, 83–84, 112, 114; as platform, 95–101, 105; and podcasts, 92; politics of, 93, 95, 100, 114; recognition and prestige of, 86, 100; and social media use, 97, 100, 108–11; Sterne on, 268; tagging and linking on, 98–100; *versus* traditional academic labor, 97–99, 104–5, 114, 115n2; writing, pushing past, 91–92
Sounding Out! Editorial Collective: about, 115n1; benefits from SO!, 99; on blogging and making, 84–85; Google Hangout screenshots, 85–87, 94, 96, 99, 113; working with writers, 97–98
Sounding Out! survey: disciplinary connections, 103–4; editorial process, 104–8; methods, 101–2; questions, 102; results overview, 102–3; social media, 108–11; writers' engagement, 111–12
sound practices: audience for, 252; and embodied listening, 251–52; goals of, 251–52, 262; immersion in digital spaces, 256–59; increasing inclusivity, 252–56; physical experiences during, 259–61
soundscapes: and audible history, 232, 234–38, 240; Feld's work on, 7; in libraries, 18n25; narrative descriptions of, 167; Projet Bretez and, 240–43; "Public Secrets" project, 258; as situated, 234; and subjectivities, 235; "A Tale of Two Soundscapes" project, 253–54; "The Roaring Twenties" project, 11, 237–238,
Sounds of British Literature website, 216
sound studies: and black studies, 5; and cultural studies, 5–6; current state of, 12, 88, 92; and digital humanities, 8, 11, 16, 48, 267, 280; and digital technologies, 7; diversity in, lack of, 93, 263n14; early history of, 86, 88; in the humanities, 168, 280; and media studies, 6; prior work in, 4–5, 7, 16n2; professional groups in, 116n13; Sterne interview on, 280–81; Western focus of, 17n6
source bias, 38
speech: paralinguistic features, 161–63, 165–67, 171; prosody, 161–63, 165–66, 168, 171; voice qualifications, 164, 169; voice qualities, 163–64
Sterne, Jonathan, interview: academic listening, lack of, 277; audio mark-up, 268, 270, 275; audio player interface metaphor, 277–78; Auto-Tune, 270, 282n1; change in the academy, 269; citing digital projects, 274; close listening, 275–76; digital humanities, 267–70, 282; digital media as visual media, 277; digital scholarship, engagement with, 274–75; distant listening, 276; on Hurston, 279; infrastructure, need for, 282; institutional problems, 270–71; labor, 271; multimedia preservation, 271;

Sterne, Jonathan, interview: (continued) music information retrieval, 275; pedagogy, 278–79; platform problems, 270–72; practice, turn towards, 269; presentations and technology, 281–82; prestige and publishing, 273–74; project maintenance, 271–72; research creation, 269; self-publishing, 272–73; sound, visual representation of, 277–78; *Sounding Out!*, 268; sound studies field, 280–81; tenure and promotion, 270, 273–75; on this book, 278–80

Systems of Prosodic and Paralinguistic Features in English, 161–64, 169

"A Tale of Two Soundscapes" project, 253–54

teaching. *See* pedagogy

technology: and audible history, 233–34; and collection accessibility, 155–56; in cultural studies, 5–6; enabling folkness, 71; in media studies, 6; Ong on, 76; and presentations, 281–82; recording equipment, 7–8; studying, importance of, 155–56

TEI's Transcriptions of Speech classifications: authenticity *versus* computational tractability, 164–65; contextual information, 170; HiPSTAS use of, 158–60; paralinguistic features in, 165–67; prosodic features in, 165–66; source for, 163; text, emphasis on, 166–67, 173n39; voice qualities and qualifications in, 163–64

tenure and promotion: and blogging, 83, 99, 115n2; and multimodal scholarship, 3, 115n2, 244; Sterne interview, 270, 273–75; writing, emphasis on, 11, 244

Text Encoding Initiative (TEI), 10, 163, 165–66

text mining, 9

"unholy trinity," 53–54
universal design, 15, 253–55

vernacular culture, 7, 65, 67, 71, 73, 76, 148n5. *See also* Berkeley Folk Music Festival; Choir! Choir! Choir!; folkness; folk revival, American

vernacular expression, 65–66
vibration, 260
Virtual Paul's Cross Project, 235–37, 240, 246n13
vocal gestures, 159
Voyage to the Islands, 29–30, 32–35, 38
Voyage to the Islands, re-creation of music from: the bamboomba, 35, 41; bias, problems created by, 38–39; cassette recording for, 35–36; digital recording for, 36–37, 41; documentarian rejection, 39–40; and historical imagination method, 32–33; instruments created and modified for, 32–35; and microtones, 35, 37–39, 41; MIDI, use of, 36–37, 41–42; Musical Passage: A Voyage to 1688 Jamaica, 43n2; music created for, 32–34, 39, 41, 43n1; music creation, importance of, 38; overview of, 32; *sansas*, 34–35, 37, 41, 43n15; slavery questions raised by, 40

Voyant Tools, 9

Watson, Arthel Lane "Doc," 182–86, 188–90

West African music, 33–34

www.ingramcontent.com/pod-product-compliance
Lightning Source LLC
Chambersburg PA
CBHW070753230426
43665CB00017B/2342